THE DISPERSED CITY:

THE CASE OF PIEDMONT, NORTH CAROLINA

Charles R. Hayes

*The University of North Carolina
at Greensboro*

THE UNIVERSITY OF CHICAGO
DEPARTMENT OF GEOGRAPHY
RESEARCH PAPER NO. 173

1976

Library of Congress Cataloging in Publication Data

Hayes, Charles R.   1919—
    The dispersed city.

    (Research paper — The University of Chicago, Department of Geography; no. 173)
    Bibliography: p. 151
    1. Cities and towns—Piedmont region.   2. Cities and town—North Carolina.
3. Piedmont region—Economic conditions.   5. Urbanization—Piedmont region.
6. Urbanization—North Carolina.   I. Title.   II. Series: Chicago. University.
Dept. of Geography. Research paper; no. 173.
H31.C514 no. 173 [HT123.5.N8]   910s
ISBN 0-89065-080-2 [301.36'3'097566]
76-16849

Research Papers are available from:
        The University of Chicago
        Department of Geography
        5828 S. University Avenue
        Chicago, Illinois 60637
    Price: $6.00 list; $5.00 series subscription

To family and friends

After a decade of living in Greensboro, North Carolina I real-
ized I would never return to the Chicago area to live, even though
I had spent most of my life there. Hard put to analyse my preference
for the North Carolina Piedmont, I thought in terms of the pleasant
climate, the scenic rolling hills, the year-round greenery, and the
friendly people. I knew Greensboro was a unit of a "dispersed city",
or thought I did, and even said so in print a couple of times. However, I assumed that dispersed city character had nothing to do with
the pleasant life. Then I read Norton Ginsburg's speculations on
The Dispersed Metropolis: The Case of Okayama and began to wonder
if, indeed, the dispersed city does make the traditional city an
anachronism in modern life. Perhaps the things I liked about living
in Greensboro were the very things that characterize a dispersed
city. It might be worth investigating. The results of the investigation are reported in the following pages and pose more questions
than they answer. Nevertheless, I believe both questions and
answers are worth reporting.

I am indebted to Norton Ginsburg for his critical review of the
manuscript. The UNC/G students who did the interviewing were necessary to the project. The UNC/G administration has been very patient
with me, and my family has not been too critical of the "midnight
oil." I guess, though, we all get by with a little help from our
friends.

TABLE OF CONTENTS

LIST OF GRAPHICS

Chapter I

INTRODUCTION

The Dispersed City Concept

A strip of urbanism from Virginia through the Carolinas to
Georgia marks the home of the Southern Piedmont Manufacturing Belt.
Known for its cotton textiles, tobacco, machinery, lumber, and
furniture products, this manufacturing belt, located on a transitional land form of rolling hills between the Appalachian Mountains
and the Atlantic Coastal Plain, is attracting manufacturing industry in many categories at a rapid rate.

The North Carolina segment of the Piedmont is a crescent-
shaped slice of urbanism from Raleigh on the east to Charlotte on
the south. Because of the crescentic shape of this band of urbanism,
it is often referred to as "The North Carolina Piedmont Crescent."
Midway between Raleigh and Charlotte, a group of six small cities
lies athwart this crescentic strip. Do these six urban nodes and
the non-urban land surrounding them constitute a "dispersed city?"

Ian Burton defines a dispersed city as a group of urban nodes
in close proximity to each other, separated by tracts of non-urban
land, and functioning as a single urban entity. According to
Burton, distances between nodes of a dispersed city are short enough

for customers to choose one of the several for higher level
shopping. Each urban node purveys convenience items locally but
specializes in higher level goods and services.[1] The term "dispersed city" is widely used to identify a group of closely spaced,
functionally interdependent urban nodes, but the literature reveals
no evidence that such functional agglomerations actually exist, or
if they do, that they are anything but aberrant elements in the
conventional hierarchical pattern of settlement in economically
developed regions.

The term "dispersed city" first appeared in the geographic
literature in 1953 and referred to a group of cities in southern
Illinois that presumably functioned as a single urban unit though
each urban node was physically separated from the others.[2] The
area in southern Illinois, containing the group of closely spaced
small cities, was a logical place for the application of the term
and concept because the development of this urban agglomeration was
evident.[3] The locations and thus the spacing of these cities were
influenced by the distribution of accessible seams of coal near the
earth's surface. As the coal resource was depleted, economic decay
set in, and it thus became necessary for each small city to change
its economic role if it was to survive at all. This may have been
accomplished through nodal specialization in some comparably high
level good or service.

---

[1]Ian Burton, " A Restatement of the Dispersed City Hypothesis", Annals of the Association of American Geographers 53,
No. 3(September, 1963): 285-289.

[2]O.W. Beimfohr, "Some Factors in the Industrial Potential
of Southern Illinois," Transactions of the Illinois State Academy
of Science 46(1953): 97-103.

[3]Ibid. pp. 97-98.

The term dispersed city has also been used to define urban
sprawl,[1] and different terms have been suggested for the concept
of "dispersed city." For example, Ginsburg used the term "dispersed
metropolis" to refer to what might be a group of functionally interdependent urban nodes on the Okayama Plain of Japan,[2] and terms such
as "oligopolis" and "poly-nucleated metropolitan region" have been
suggested informally although not in the published literature.
Burton used the term "dispersed city" in 1959 and again in 1963 to
refer once more to the southern Illinois group of cities.[3] Berry,
Mayer, et al. used the term in 1962 to refer to the concept of closely-
spaced functionally interdependent urban nodes,[4] and Berry used the
term again in 1967 to identify the same sort of urban agglomeration.[5]
The term "dispersed city" seems the most useful phrase with which
to identify this type of urban form when it is believed to exist.

Although the term seems well enough established, the
general proposition that underlies it remains inadequately substantiated. The proposition suggests that a group of closely spaced but
spatially and administratively discrete cities can function as a
single economic unit. Burton suggests that dispersed city interdependence is reflected most strongly in retail trade patterns and
that retail specialization appears for such items as furniture,

[1]Allen K. Philbrick, Analysis of the Geographical Patterns
of Gross Land Use and Changes in Numbers of Structures in Relation
to Major Highways in the Lower Half of the Lower Peninsula of
Michigan (East Lansing: Michigan State University, 1961).

[2]Norton S. Ginsburg, "The Dispersed Metropolis: The Case
of Okayama, " The Toshi Mondai, Tokyo, 52, No. 6 (1961): 631-640.
(In Japanese)--(English translation available from N.S.Ginsburg,
13 pages).

[3]Ian Burton, "Retail Trade in a Dispersed City," Transactions of the Illinois State Academy of Science 52 (1959): 145-50
and I. Burton, "A Restatement of the Dispersed City Hypothesis."

[4]B. J. L. Berry and H.M.Mayer et al., Comparative Studies
of Central Place Systems, Report to: Geography Branch, Office of
Naval Research (Chicago: Department of Geography, University of
Chicago, 1962), part IV, pp. 6-9, and Fig. 1.

[5]Brian J. L. Berry, Geography of Market Centers and Retail
Distribution (Englewood Cliffs, N. J.:Prentice Hall, 1967),
pp. 34 and 58.

clothing, footwear, automobiles, and radio and television sales and
service. Functional specialization patterns, according to Burton,
may be reflected in retail sales returns and in traffic flow.[1]
Burton suggests that a clue to the existence of dispersed cities
is population size. No predominant urban node of a dispersed city
will have a population twice that of its nearest rival, and several
cities will be in the same size class of population.[2] Using this
clue, Burton has identified dispersed cities in Ontario and Texas
in addition to the southern Illinois agglomeration.[3]

Berry evidently accepts the definition offered by Burton
and suggests that if the nodes of a dispersed city are considered
as a single entity, the resulting urban unit fits the central-place
hierarchy and lies between suburban shopping districts and rural
area market towns with regard to trade area size and trade area
population density. However, according to Berry, if each node of
a dispersed city is considered a separate urban entity, the individual nodes do not fit the central-place hierarchy of urban places.[4]

Ginsburg describes the "dispersed metropolis" in the original
meaning of the term "dispersed city" as a system of sub-centers
separated by considerable non-urban space but joined into an opera-
tional entity -- a metropolis without a core. He suggests that the
several cities of the Okayama plain -- Okayama, Kurashiki, Kojima,
Tamano, Tamashima, Saidaiji, and Soja -- might fit the dispersed
urban entity model. The Ginsburg model is speculative and really
proposes "what might be" rather than "what is." Nevertheless,
Ginsburg, in a neo-Weberian fashion, offers seven characteristics

---

[1]Burton, "A Restatement of the Dispersed City Hypothesis,"
p. 286.

[2]Ibid. p. 286.

[3]Ibid. pp. 287-288.

[4]Berry, p. 34.

of such an "ideal-typical" dispersed metropolitan system.[1]

1) It would be composed of several built-up areas. . . separated from each other by . . . open space, in . . . non-urban land uses.

2) Each unit should have roughly the same population size, although differences in population will reflect the nature of the functions each unit performs.

3) Although each unit would possess certain common ubiquitous functions, such as those associated with the provision of day-to-day necessities and retail services, each would have some major specialized activity -- certain types of manufacturing, port functions, wholesaling, major retail shopping, higher education, specialized entertainment, medical services, and the like.

4) The system could have a common government. . . as would be the case in the conventional metropolis.

5) Development might be ordered by means of a general development plan. This would include effective zoning in the interstitial areas as well as in the built-up units themselves. In this way open spaces could be preserved, and certain of the drawbacks of the conventional crowded metropolis would be minimized.

6) Each unit would be linked by a system of communications as complete and techonologically advanced as science and economy could provide.

7) The . . . population in the interstitial open areas probably should have agriculture as one of its major economic bases, but . . . would be so well integrated into metropolitan life . . . that (they) could best be called "urbanized farms."

Ginsburg sees the communications system as the crux of the matter. If a resident lived in one urban unit, worked in another, obtained legal services in yet another, and so on, overcoming the

---

[1]Ginsburg, pp. 6-8 (English translation).

friction of distance would pose a considerable problem. Further, Ginsburg believes that an advanced transport and communications system is necessary if the "Idea of the City" is to be developed among its inhabitants. A city, according to Ginsburg, is an idea or concept as well as a physical object. If a "dispersed metropolis" were to conform to these several criteria, it may be regarded as the optimum urban environment of the future.[1]

The Piedmont North Carolina Urban Complex

The dispersed city form of urbanization has been tentatively identified in southern Illinois, Ontario, Texas, and southern Okayama prefecture in Japan. The author of this study also has tentatively identified two such agglomerations in North Carolina. These units have been described in terms of their downtown trade area and employment field overlap.[2] 1) The North Carolina Coastal Plain Dispersed City said to consist of five closely spaced urban nodes that seem, according to trade area and employment field data, to function as a single urban unit. 2) The so-called North Carolina Piedmont Dispersed City is an urban complex consisting of six closely spaced urban nodes that also seem to function as a single urban system. The second is the object of this study.

---

[1] Ibid. pp. 3-4 and pp. 7-8 (English translation). "The city may be regarded in yet another but related way. It may be described as an 'Area of Maximum Spatial Interaction,' The place in which more people, ideas, and things are brought together than any other place. Spatial interaction may well be the most convenient single concept in explaining urbanization and the growth of cities. Functions and activities tend to pile one upon the other in as small a place as possible in order to minimize the function of distance and maximize the amount of spatial interaction possible. This concentration incidentally leads to the citizen's awareness of the city as an environmental entity and to the development of what might be called the "Idea of the City," for a city is an idea or concept as well as a physical object. The greater the degree of concentration, the greater the amount of interaction theoretically possible, and the greater the degree of "urbanness."

[2] Charles R. Hayes and D. Gordon Bennett, Factors of Spatial Interaction in North Carolina (Raleigh, N. C.: North Carolina Department of Adminstration, State Planning Task Force, Report No. 64.13, April 1969), pp. 1-49.

The urban agglomeration identified as the "Piedmont Dispersed City" consists of the urban nodes of Winston-Salem, Greensboro, High Point, Burlington, Asheboro, and Lexington, North Carolina, contained within the counties of Forsyth, Guilford, Alamance, Randolph, and Davidson. The six nodes contained 405,943 people in 1970 with an additional 367,160 people within five counties but beyond city boundaries.[1] Nearly eight hundred thousand people would seem to qualify for status as metropolitan residents, though they may live, work, and shop at various places in five counties, provided that the multi-nodal system of which they are a part displays characteristics of functional integration. The problem is to define the bases for that integration.

The dispersed city is described morphologically as consisting of closely spaced but physically discrete urban nodes of "roughly" the same population size. If such discrete nodes do in fact constitute an urban entity, the difference in form from the single-centered city will be obvious even to the casual observer. In the conventional city, concentration and centrality would seem to be the basic requirements for urbanity. In the dispersed city it must be demonstrated that urban functions can be separated from each other by non-urban space and at the same time integrated as an urban unit, if the hypothesis of a multi-nodal integration system is to be accepted.

The Piedmont urban complex clearly fits the morphological criteria suggested by Burton and Ginsburg. Distances between its nodes are short, although the units are physically discrete and separated by land in non-urban uses. The average distance between centers is twenty-two miles. "Roughly the same population size"

---

[1] United States Bureau of the Census, Census of Population, 1970.

(for individual units) and "several cities in the same size class
of population" are dispersed city characteristics suggested by
Ginsburg and Burton respectively. Both qualifications are subjective,
but the Piedmont urban complex can be construed to fit either or both.
"No predominant city with population twice that of its nearest rival"
is an additional Burton qualification. Winston-Salem and Greensboro
are the two largest units of the complex and so near the same population size that each claims to be the "largest" following every annex-
ation. The Piedmont agglomeration qualifies for dispersed city status
according to form. If functional interdependence also can be demonstrated, then it must be concluded that a "dispersed city" does indeed
exist in the form of the Piedmont agglomeration.

Purposes of the Investigation

Is the degree of interaction between nodes of the presumed
Piedmont Dispersed City sufficient to consider the agglomeration a
single functional entity; that is, do residents live in one node, shop
in another, work in a third, seek higher education in yet another, and
so on? Chapter II attempts to answer these questions. That chapter
examines the function of the Piedmont complex through consideration
of various forms of interaction among the six urban nodes. Shopping
patterns, work patterns, and media consumption patterns through service area delineation are the basis for inferring dispersed city function in the chapter. If it can be inferred through service area
delineation that residents of the complex often live in one node,
shop in another, work in a third, and consume media transmission from
yet another urban node, inter-nodal interaction and specialization
are implicit, and the hypothesis that a dispersed city exists will
have been demonstrated.

If the hypothesis that a dispersed city exists in the
North Carolina Piedmont is accepted, the obvious next question is why
does the phenomenon appear there? How could an urban region develop

without the concentration and centrality usually associated with the conventional urban form? Burton, as does Ginsburg, suggests that the emergence of such centers depends, at least in part, upon the level of transport technology operating in the formative stage of development.[1] Even transportation, however, can be only one explanatory variable.

Chapter III considers these questions in an historical context. Of course, spacing can tell only part of the story of dispersed city development. In addition, each node must have grown at about the same rate from inception to the present or one node would have become dominant over time. A dominant node surrounded by satellites would not suit the definition of a dispersed city since it is characteristic of conventional forms of urbanization, as in the Chicago metropolitan region, for example. Chapter III discusses some of the reasons for the fairly even growth rates of the various nodes in the Piedmont urban complex after briefly tracing the historical development of the five countries in the area.

Chapter IV contains an analysis of population numbers and distribution in the area and a review of the efforts at population planning for the multi-nodal urban region. The control of population growth and distribution--or lack of control-- can have a profound effect on the form and function of the Piedmont Dispersed City (if it does in fact exist) for future generations. If the dispersed city form of urbanism is to exist in the twenty-first century, population growth and distribution must be considered.

If the dispersed city exists, to what extent are the residents aware of the fact? Is the "idea of the city" developed among its inhabitants? Do the residents of the Piedmont area consider each separate urban node a city unto itself, or do they recognize the

---

[1]Burton, "A Restatement, etc." op. cit. p. 286 and Ginsburg, op. cit. p. 3.

agglomeration of nodes as a single urban entity? Ginsburg asks
these questions about resident perception of the dispersed city.[1]
If, as Ginsburg suggests, a city is a concept as well as a physical
entity, these are indeed important questions. Chapter V deals with
these questions. Chapter V attempts also to assess the resident's
perception of the urban area in this and in other impressions. Do
residents perceive nodal specialization? Do residents like living
in what appears to be a decentralized urban system? Why or why not?
Do they want a common government? If so, what do they think its
functional characteristics should be? What do they like or dislike
about present government services, shopping services, professional
services, travel, and so forth?

Chapter V attempts to discover what the Piedmont Dispersed
City "seems to be" to its residents on the assumption that their
behavior will reflect their attitudes and perceptions. The phenomenological approach to environmental perception assumes that it does
not matter what the situation really is, only what it seems to be.[2]
For example, if a person believes shopping place A purveys higher
quality merchandise than shopping place B but believes prices to
be identical at both places, all else being equal, he will likely
behave in accordance with this perception of the situation even
though it may be in error. In fact, considerable retail advertising
is directed toward erasing or re-inforcing perceptual error. In
this case quantifiable data can be applied toward perception.
Suppose, however, a person believes a common government to be desirable for dispersed city administration. He may be right or wrong,
but he will probably vote in accordance with his perception of the

---

[1]Ginsburg, op. cit. p. 4 and p. 11

[2]This theme is present in much of the geographic literature
on "environmental perception." See Chapter V for references to geographic and other literature involving perception and phenomenology.

situation. It is possible that many people live in the shadow of
perceptual error much of their lives and behave in response to "what
seems to be" instead of "what is." Chapter V, as are most geographic
perception studies, is based on interviews. A description of the
interview structure is included in the chapter and the questionnaire.
is reproduced in the appendix.

One additional hypothesis relating to the dispersed city is
examined in Chapter V, that is, do residents of such areas leave
them for top-level shopping?[1] It is possible to answer this question
by asking respondents to recall the number of trips made out of the
area during the past year and the purpose of each trip. (Since this
information was obtained through interviews, it is included in Chapter V rather than elsewhere, although it is not a study of perception per se.) The information obtained is compared to United States
Department of Labor projections for the year 1975.[2]

Chapter VI consolidates the conclusions derived from Chap-
ters I through V and suggests planning possibilities for the urban
region. The chapter also considers whether or not the "Idea of the
City" is a part of the cognition of the area resident, for although
the elements and attributes of a "dispersed city" may be present,
it can never operate in reality unless there is a consensus about
its existence.

---

[1] Burton, "A Restatement, etc." op. cit. p. 288.

[2] United States Department of Labor, Bureau of Labor Statistics, National Planning Association, ORRRC Study Report 23
(Washington, D.C. 1962), pp. 111-115.

Chapter II

INTERDEPENDENCE WITHIN THE PIEDMONT URBAN COMPLEX

"The term 'dispersed city' remains a useful
phrase by which to identify the concept of functionally interdependent cities, located in close
proximity, but physically separated by non-urban
land. It is on the established value of this concept that the utility of the term seems to rest."[1]

Introduction

The statement above by Ian Burton implies that the concept of closely-spaced but discrete urban nodes functioning as a
single urban entity is well established by empirical studies, but
in fact, although the term is widely used, there are no studies
which prove the assertion that such functional agglomerations
actually exist. The functional interdependence of the urban agglomerations identified by Beimfohr, Burton, Ginsburg, and the author
has not yet been demonstrated, only suggested. We do not really
know whether the residents of the identified dispersed cities of
southern Illinois, Ontario, Texas, Japan, and North Carolina actually live in one node, shop in another, obtain professional services

---

[1]Ian Burton, "A Restatement, etc.", op. cit. p. 289.

in a third, work in yet another, and so on. However, if functional
interdependence of a multi-nodal urban system of the sort described
in Chapter I can be demonstrated in even one major instance, the
concept of the dispersed city as a geographical phenomenon must be
considered considerably more plausible than it now is. The purpose
of this chapter is to explore the extent of that interdependence in
the case of the Piedmont urban complex. It has been pointed out that
the urban nodes of Winston-Salem, Greensboro, High Point, Burlington,
Asheboro, and Lexington, contained within the counties of Forsyth,
Guilford, Alamance, Randolph, and Davidson and located on the North
Carolina Piedmont Crescent midway between Raleigh and Charlotte,
approximate the model for dispersed city form proposed by Burton
and Ginsburg, but their functional interdependence remains to be
demonstrated.

There are several ways to solve the problem of demonstra-
ting functional interdependence between nodes of a dispersed city.
Burton suggests traffic flow and retail sales returns.[1]

Traffic-flow measurements do, of course, indicate movement
between nodes. Note on Map 1, adapted from the North Carolina State
Highway Commission state traffic flow map for 1970, that there is,
indeed, considerable automobile traffic flow between nodes of the
Piedmont urban complex. There is also, however, considerable through
traffic along the arc of the North Carolina Piedmont crescent as a
whole. The pattern of the map is indicative, not definitive, and
it requires further elaboration.

Published retail sales data cannot be used to demonstrate
functional interdependence because census data on retail sales do
not designate the destination of a retail item. Valuable as this
information is for certain purposes, it does not help solve

---

[1]Ibid. p. 286.

the problem proposed here. Nevertheless, shopping is such an important potential interaction generator that it is considered in detail in this chapter through service-area delineation and by cartographic plotting of interviews.

If a person lives in one node, shops in another, works in a third, seeks entertainment in another, and sends his sons and daughters to college in yet another node of a dispersed city, this will be reflected in the service areas or ranges of these various functions. For example, central business district trade-area delineation will show where people live, who shop in a specific central business district. Employment field delineation will designate where residents work who live in a certain urban node. Media Service areas -- newspaper, radio, and television -- will indicate where residents live who consume these forms of entertainment. A commuting service area for one of the institutions of higher learning within the area will designate where the young residents live who attend that institution.

Service Area Delineation

The technique used to delineate service areas requires some explanation. It was developed by the author through considerable trial and error over a period of several years. The technique evolved from many consulting efforts to accurately delineate service areas for retail stores, shopping centers, banks, newspaper publishers, television broadcasters, and other functions that serve customers within a given area. The owners and managers of these functions require a cartographic portrayal of service area plus a numerical designation of service-area size. The cartographic representation permits analysis of population characteristics within the · service area, and the numerical designation permits size comparison with other service areas. The technique used here has been refined through repeated efforts, until both cartographic and numerical designations are quite accurate. It has been possible to establish the accuracy of the method by comparing the results of the sampling technique described below with known total populations. However, since it is only very rarely possible to obtain data on total populations, sampling techniques must be used in most cases.

The first step in service-area delineation is to collect on-location interviews. Interviewers stand at check-out lanes or at any convenient location where the flow of retail customers can be interrupted for questioning. For service-area delineation it is necessary only to obtain residence addresses, although other information can be collected at the same time. Applebaum and Spears claim that randomness is obtained by questioning each customer who passes a specific point. The interviewer secures best results by interviewing at this point for an hour or two, two or three times per day.[1] There is certainly no harm in interviewing in this fashion, but sampling accuracy is no greater than that obtained through any interviewing technique. For example, the trade area for Greensboro's central business district was measured and delineated four times over a period of five years. Each successive delineation was almost identical to the last, yet different interview techniques were used each time. Further, it matters little how many interviews beyond about one hundred are collected. Two hundred, or even five hundred, interviews do not increase the accuracy of service-area delineation sig-

NB

---

[1] William Applebaum and Richard Spears, "Controlled Experimentation in Marketing Research", The Journal of Marketing (Jan. 1950), pp. 505-517 and Idem, "How to Measure a Trading Area", Chain Store Age (Jan. 1951), revised brochure, no pp. nos.

MAP 1 - 1970 PRIMARY HIGHWAY SYSTEM TRAFFIC MAP

The map depicts the average 24-hour traffic flow for all
vehicles on the designated primary highways within the central
North Carolina Piedmont including the five-country study area.
Primary highways include interstates, U. S. highways, and N. C.
highways. Although there is considerable automobile traffic
within the study area, there also is considerable traffic outside
the area especially along the arc of Interstate 85 from Charlotte
to Raleigh.

17

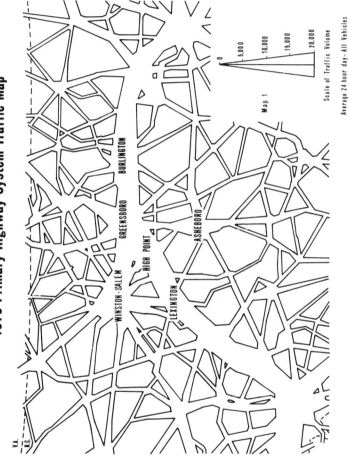

1970 Primary Highway System Traffic Map

WINSTON-SALEM

GREENSBORO

BURLINGTON

HIGH POINT

LEXINGTON

ASHEBORO

V.A.
N.C.

Map 1

Scale of Traffic Volume

0
5,000
10,000
15,000
20,000

Average 24 hour day – All Vehicles

nificantly.[2]

The second step in service-area delineation is to plot residence site samples on a map, measure the distances from each site
to the function, and compute mean and standard deviation for the
distances.

The third step in service-area delineation is to compute
one-tailed ninety-nine per cent confidence limits or confidence
intervals (as they are variously termed) of sample distance to function and compute the center of gravity of the plotted samples.

The fourth step in service-area delineation is to draw a
circle on the map, from the center of gravity of the plotted sample
distances, with the circle radius equal to the ninety-nine per cent
confidence limit of the plotted samples.

The final step in service-area delineation is to contract
the circle circumference (inward) in the areas within the circle
where no sample dots appear, and to extend the circle circumference
outward so as to include areas beyond the original circumference
that contain several sample dots. The (outward) extensions

---

[2]See, (1) Charles R. Hayes and Norman Schul, Greensboro
Retail Core Analysis (Greensboro, City of Greensboro, Planning
Department, March 1965)pp. 1-30; (2) Charles R. Hayes, "Greensboro's Downtown Trade Area," Planning Notes (Greensboro Planning
Department, City of Greensboro, 12 July 1967) pp. 1-4; and (3)
Charles R. Hayes and D. Gordon Bennett, Factors of Spatial Interaction in North Carolina, Report 64.13(Raleigh, State Planning
Task Force, North Carolina Department of Administration, April,
1969) pp. 1-70. The 1965 report was based on 359 interviews,
the 1967 report on 90 interviews, and the 1969 report on 120
interviews. (4) The fourth Greensboro central business district
trade area delineation was done in 1970 for a private client and
is, therefore, confidential, but delineation was accomplished
with 200 interviews. There were no significant differences in the
four efforts at the five percent significant level. Further,
interview techniques were not identical for each effort. Delineations one and two were accomplished through the Applebaum and
Spears technique. Delineation three was accomplished by interviewers standing outside the stores for one full day. Delineation
four was also completed outside the stores but for periods of only
twenty or thirty minutes for a three- or four-day period. Interviews for delineations one and two were obtained by questioning
each customer who passed a specific point. Interviews for delineations three and four were obtained from anyone who would answer
the questions.

must balance the (inward) contractions in order to retain the original circle circumference area. The resulting isarithm delineates the service area.

Retail service areas vary in size by day of the week and season of the year. According to the day of the week, retail service areas are largest on Friday and Saturday, average on Wednesday and Thursday, and smallest on Monday and Tuesday. Where retail functions are open for business on Monday evening and Sunday afternoon, the Monday evening service area equals the week-end service area, and the Sunday afternoon service area is largest of all. According to season of the year, retail service areas are larger than normal between Thanksgiving and Christmas and for the month preceding Easter. Special events also extend the retail service areas. For example, a major golf tournament is scheduled in the Piedmont urban complex the first week in April of each year. While this tournament is in progress, all retail service areas are larger than normal for day of the week and time of the year.

Retail service areas also vary in size by retail function. If an investigator is measuring and delineating the service area for a group of retail functions, such as those in a central business district or a shopping center, this becomes important. Department stores, new car sales showrooms, office supply stores, and clothing stores have the largest service areas. Auto accessory stores, jewelry stores, furniture stores, and variety stores, have medium-sized service areas. Small appliance stores, hardware stores, loan agencies, and drugstores have small service areas.[1] This will come as no surprise to any student of marketing or of "central-place theory", which postulates a hierarchy of retail functions each with different threshold requirements and different ranges for the goods or services

---

[1]Hayes and Schul, op. cit. p. 4.

purveyed.[1]

If an investigator wished to delineate the maximum extent
of a retail service area (termed "economic reach"), day of the week,
season of the year, special events, and type of retail function must
be considered. Service areas delineated in this chapter are mid-
week and summer, with a cross-section of functions. Service
area reach here is, then, an average of these temporal and categori-
cal variations.

Retail Service Areas

Retail shopping has been proposed by all previous investi-
gators of dispersed city function as a potential inter-nodal, interaction-generating function. Burton suggests the interdependence
of the urban places in a dispersed city is reflected most strongly
in their retail trade patterns.[2] Berry defines dispersed cities in
terms of shopping patterns and trade-area density,[3] and Ginsburg
mentions shopping as one of several interaction-generating functions.
Shopping, therefore, will be the first potential interaction-
generating function considered here.

Map 2 delimits the central business district trade areas
of the six major urban nodes of the Piedmont urban complex. Although
there are a few regional shopping centers (that is, non-CBD)
scattered throughout the region, the six central business districts
function as regional shopping centers for the area.[4]

---

[1]See, for example, Berry, op. cit. p. 15.

[2]Burton, "A Restatement . . . etc.", op. cit. p. 286.

[3]Berry, op. cit. p. 34.

[4]See Hayes and Bennett, op. cit.; Charles R. Hayes and
Normal Schul, Greensboro Shopping Center Trade Areas (Greens-
boro, City of Greensboro, N.C. May 1964); Charles R. Hayes and
Norman Schul, Greensboro Retail Core Analysis (Planning Department,
Greensboro, N.C. March 1975); and Charles R. Hayes and Norman
Schul, "Some Characteristics of Shopping Centers", Professional
Geographer, Vol. XVII, No. 6 (Nov. 1965) pp. 11-14.

Trade-area data were obtained through interviews with shoppers in the downtown districts of the nodes of the Piedmont complex. Trade-area delineation was derived through use of the method described previously.

It is evident from an examination of Map 2 that there is considerable downtown trade-area overlap in the area. A shopper almost anywhere within the urban complex avails himself of at least two central business districts for shopping purposes. Some shoppers exercise as many as four choices of downtown shopping place. Shopping interaction among nodes of the Piedmont urban complex is pronounced, as implied by a fifty-one per cent trade-area overlap. Shopping patterns implied by trade-areas present the region as one retail market area rather than a group of discrete trade areas.

Moreover, the possibility that the Piedmont urban complex is a single market area is recognized by many local business men charged with marketing a good or a service. The author has discussed this possibility with radio, television, newspaper, and retail executives on various occasions, and all have stated that they have long considered the area one market. All believe that the potential customer perceives inter-nodal retail specialization, and all credit advertising with this perception. These executives, however, also believe that inter-nodal specialization does, in fact, exist in certain top-level retail functions within the area. Unfortunately, there is no published evidence to support or refute this proposition: There is little question that the perception of retail specialization is widespread, but does it actually exist in the Piedmont complex? Surely not, as Burton suggests in furniture, clothing, footwear, automobiles, and radio and television sales and service. These goods and services are equally available in all nodes in a plethora of

MAP 2 - PIEDMONT URBAN AGGLOMERATION:
CENTRAL BUSINESS DISTRICT TRADE AREAS

The map depicts the area within which 99 per cent of the
customers live, or more technically the 99 per cent confidence
intervals (derived statistically). These confidence intervals
represent a reasonable approximation of central business district
trade areas for the six major urban nodes of the complex. The
central business district trade areas of Burlington and Lexington
encompass little more than the counties within which the cities
are located. The High Point central business district trade area
is much smaller than all the others. This is probably partly due
to High Point's restricted location and partly due to the "negative
image" (Chapter V) projected by the High Point central business
district. The Asheboro central business district trade area extends
somewhat beyond county boundaries in all directions and is comparable
in size to the Greensboro and Winston-Salem central business district
trade areas, even though Asheboro is smaller in population than are
Greensboro and Winston-Salem. The Greensboro central business district trade area includes the county within which Greensboro is
located and portions of surrounding counties, and extends north-
east (of the map) to encompass areas in that direction. The Winston-
Salem central business district trade area includes the county in
which Winston-Salem is located and portions of surrounding counties,
and extends northwest (of the map) to encompass areas in that direction. Total trade area overlap is about 51 per cent.

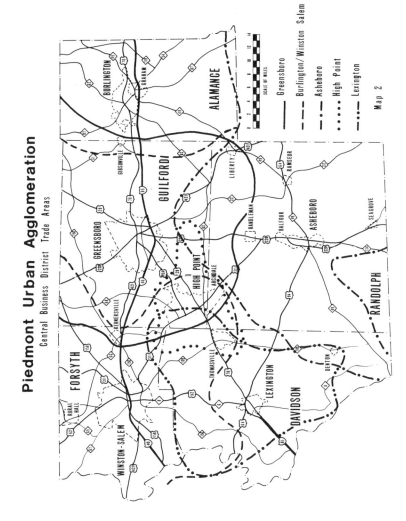

Piedmont Urban Agglomeration

Central Business District Trade Areas

Map 2

brands and styles. It is equally likely that retail specialization does not exist, in actuality, in the more esoteric retail offerings Why then the perception of retail specialization? This perception is probably based on "image". Chapter V suggests that various retail clusters project different images to the residents of the area. Each conveys a "feeling of uniqueness" so it is possible that this perception of uniqueness results in the perception of specialization.

Maps 3 and 4 indicate cross-commuting for downtown shopping purposes in a different manner than does Map 2. Maps 3 and 4 were compiled from in-home interviews with two hundred residents of the area selected at random.[1] Each respondent was asked in which downtown he shopped most often and how often and in which downtown he shopped second most often and how often. Map 3 indicates the first choice of each respondent, and Map 4 the second choice. A tally of answers to the questions concerning first and second downtown shopping choice revealed that both choices involved about two trips per month; that is, there was no difference in frequency of downtown shopping visits between the two central business districts. Further, all respondents had at least two shopping choices. However, trip frequency for third downtown shopping choice declined considerally. Only about two-thirds of the respondents had a third choice, and the third choice was visited only about once every other month.

Maps 3 and 4 indicate considerable cross-commuting for shopping purposes. This cross-commuting relates to the hypothesis that residents of a dispersed city often live in one node and shop in another. The entire sample of respondents shopped in at least one central business district other than that of their residence,

---

[1]The sample structure is explained in Chapter V and the questionnaire is reproduced in the appendix.

MAPS 3 and 4 - PIEDMONT URBAN AGGLOMERATION:
FIRST CBD SHOPPING CHOICE AND
SECOND CBD SHOPPING CHOICE

The maps indicate first and second central business district shopping choices for the two hundred-person, spatially random, sample of the entire, five-county study area (Chapter V). Since both first and second shopping choices involve an average of two trips per month, realistic trade areas surrounding each node would fall somewhere between the close-by trade areas suggested by Map 3 and the farther reach of the trade areas suggested by Map 4. This is so because on any given day, some of the people might be visiting choice one, while others visit choice two. Map 2 represents this average situation. A further difference between Maps 3 and 4 and Map 2 is the fact that Maps 3 and 4 cannot extend beyond study-area boundaries, whereas the data for Map 2 were collected state-wide. Although, taken together, Maps 3 and 4 indicate considerable cross-commuting for shopping purposes, the cross-commuting is seldom at variance with distance: that is, although the second shopping choice involves travel for greater distances than does the first shopping choice, almost all second choices involve the next closest central business district. Further, a comparison of Maps 3 and 4 with Map 1- reveals that implied travel patterns follow the "primary" highway system rather closely.

Piedmont Urban Agglomeration

First CBD Shopping Choice

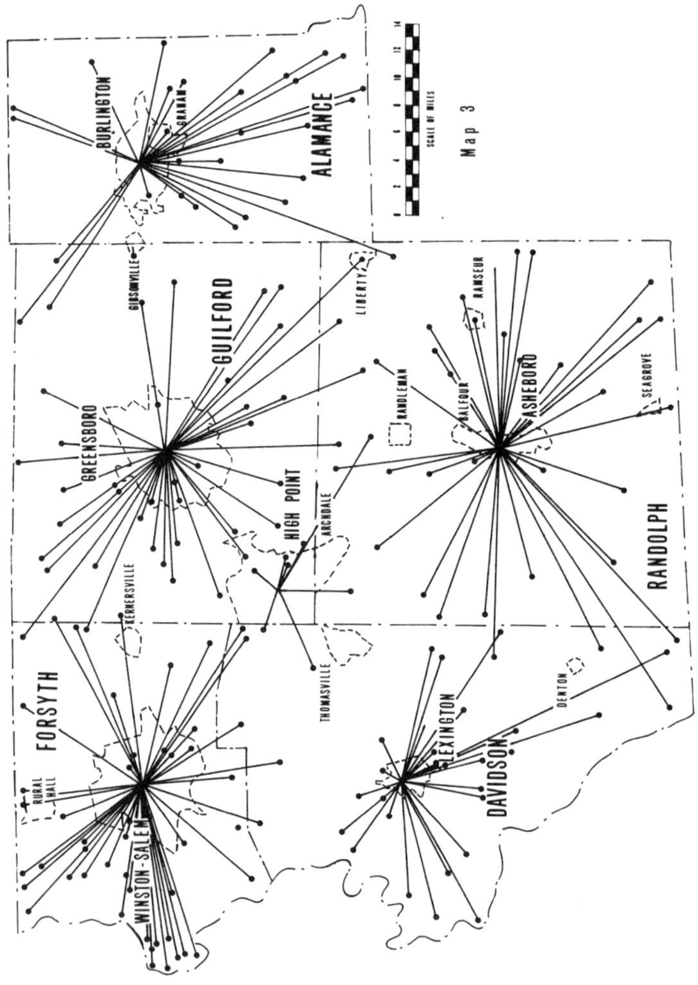

Map 3

Piedmont Urban Agglomeration

Second C B D Shopping Choice

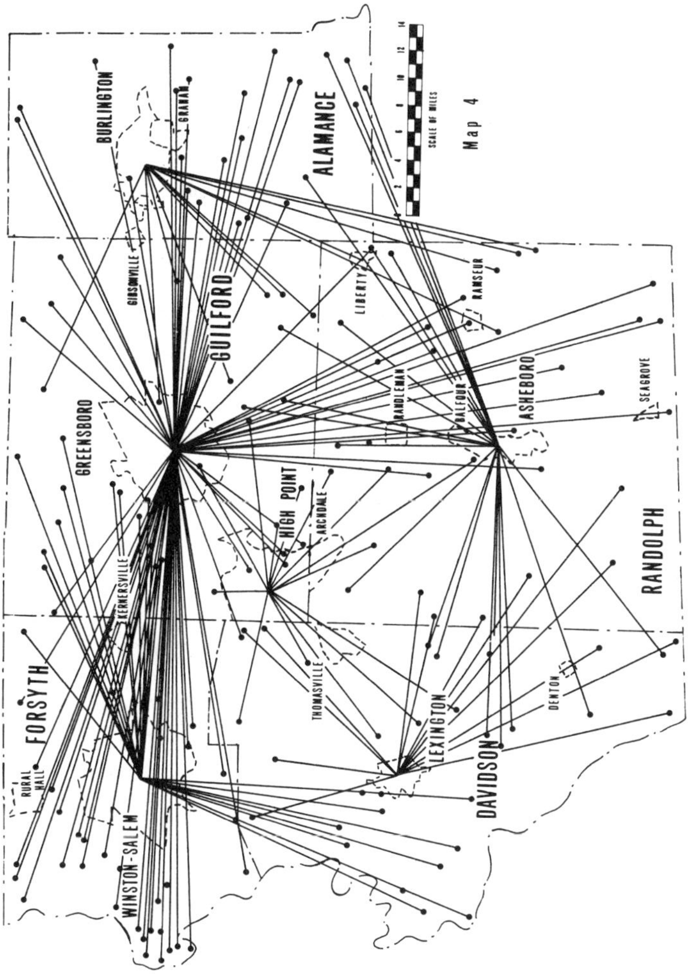

BURLINGTON
GRAHAM
ALAMANCE
GIBSONVILLE
GUILFORD
LIBERTY
RAMSEUR
RANDLEMAN
BALFOUR
ASHEBORO
SEAGROVE
GREENSBORO
HIGH POINT
ARCHDALE
RANDOLPH
KERNERSVILLE
FORSYTH
RURAL HALL
THOMASVILLE
LEXINGTON
DENTON
DAVIDSON
WINSTON-SALEM
SCALE OF MILES
0 2 4 6 8 10 12 14

Map 4

and sixty-six per cent of the sample had a third downtown choice, although the third choice was not visited with the frequency accorded choices one and two. Further tabulation reveals that about 10 per cent of the sampled residents even visit a fourth choice for downtown shopping but only infrequently.

Although Maps 3 and 4 indicate considerable cross-commuting for shopping purposes, it is interesting that this cross-commuting is seldom at odds with distance friction. Although second shopping choice involves travel for greater distances than does first shopping choice, almost all second choices involve the next closest downtown. Further, implied travel patterns follow the four-lane highway network rather closely. The private automobile is the prime transporter of people within the Piedmont urban complex. There is no rail rapid transit in the region, and even bus service is meager.

As stated, Maps 2, 3, and 4 imply considerable cross-commuting for shopping purposes within the area. Residents living in a specific urban nodes or in the non-urban spaces close to a specific node often shop in the central business district of a different node. Although this does not prove inter-nodal retail specialization, it does suggest as has been pointed out, that residents believe that internodal retail specialization exists within the urban complex.

It has been pointed out that the Piedmont urban complex functions as a single market area in terms of retail shopping habits. This concept is supported by the pattern of television-viewing within the area. Three commercial television stations serving the Piedmont complex are located in three separate urban nodes: one in Greensboro, one in Winston-Salem, and one in High Point. The sending towers are even farther from each other than are the cores of the three urban nodes. Potential signal range would indicate that the stations serve

three separate cities; that is, some people who can receive one of
the signals cannot receive the other two. The signals sent do not
cover exactly the same area. Nevertheless, the three signals are
consumed in very nearly the same area. Signal consumption overlap
is eighty per cent. Further, nearly all the television viewers of
the Piedmont urban agglomeration watch all three stations at some
time during the day or week. Very few television viewers watch one
and not the other two stations. The three stations might as well
be in a single urban node for they function as if they covered a
single market area. Further, the service areas of the three stations
are very near the same size, although the potential signal ranges
differ in size for the three stations because of signal frequency,
tower location, and terrain. Measured by the 99 per cent confidence
limits of consumption, the television service areas of the three
stations are as follows: Greensboro 57.8 miles; Winston-Salem 65.8
miles; and High Point 61.0 miles. These numbers are not significantly
different at the 5 per cent significance level, nor are the means
and standard deviations of signal consumption distance from which
confidence limits were derived. Map 5 shows the spatial similarity
of viewer location. Thus, the pattern of television consumption
strongly suggests that the area is a single market.

Map 6 delineates the radio service areas for these major
urban nodes. The several radio stations of each node have been
combined on the map to appear as if they were one station broadcasting
from each urban node. Map 6 shows that a resident anywhere within
the area not only can, but does, avail himself of the radio signals
emanating from the other nodes of the Piedmont complex. The Greensboro signals are received by some residents of all six major nodes
plus Chapel Hill to the east of the area. Winston-Salem radio signals
capture listeners in High Point and Lexington, beyond local city

MAP 5 - TELEVISION SIGNAL CONSUMPTION

The map depicts the 99 per cent confidence intervals of
television signal consumption of the three stations serving the
Piedmont urban agglomeration. The lines on the map represent the
areas where 99 per cent of the viewers reside. This differs from
the signal reach which, because of signal frequency, tower location,
terrain, and cablevision, might be greater or smaller than the consumption areas. Although consumption overlap is 80 per cent, consumption areas are not identical. Channel 12 in Winston-Salem
captures viewers farthest to the west, Channel 8 in High Point
captures viewers farther west than does Channel 2 in Greensboro, but
but not as far as does Channel 12. The pattern of viewer consumption is reversed to the east of the study area. Channel 2 in
Greensboro reaches further to the east for viewers; Channel 8 is
next and Channel 12 least. All three stations capture viewers
from the same areas to the north and south of the study area.
Almost all television viewers of the Piedmont urban agglomeration
watch all three stations during a week. As far as the study area
is concerned, the signals from the three stations are consumed
as if they emanated from a single city. The pattern of consumption
suggests that the study area is a single market.

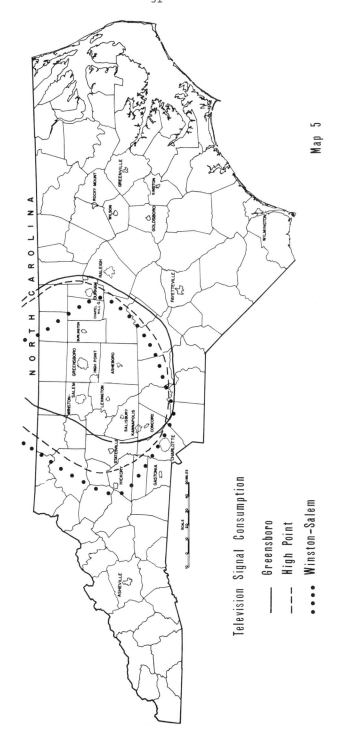

Television Signal Consumption

——— Greensboro

– – – High Point

•••• Winston–Salem

Map 5

MAP 6 - RADIO SIGNAL CONSUMPTION

The map depicts the combined 99 per cent confidence intervals
of radio signal consumption for all radio stations broadcasting from
the six major urban nodes within the study area; that is, the consumption areas have been combined on the map to appear as one station
broadcasting from each of the major nodes when, in actuality, each
consumption area represents several radio stations. Signal consumption overlap is forty-six per cent, which is considerably less
than television signal consumption overlap. In fact, the radio
signals emanating from Winston-Salem, Burlington, Asheboro, and
Lexington are not consumed far outside the county boundary within
which those cities are located. (Data for High Point radio signal
consumption are not available.)   Only the Greensboro radio signals
are consumed by listeners from within most of the study area. The
Greensboro radio signals are not more powerful, just more popular.
They are consumed as if the study area was a single market. The
radio signals emanating from the other urban nodes of the study area
are consumed as if each node served just its surrounding county.
Nevertheless, the consumption overlap and the popularity of the
Greensboro stations suggest the one market concept, though not as
strongly as does television signal consumption.

33

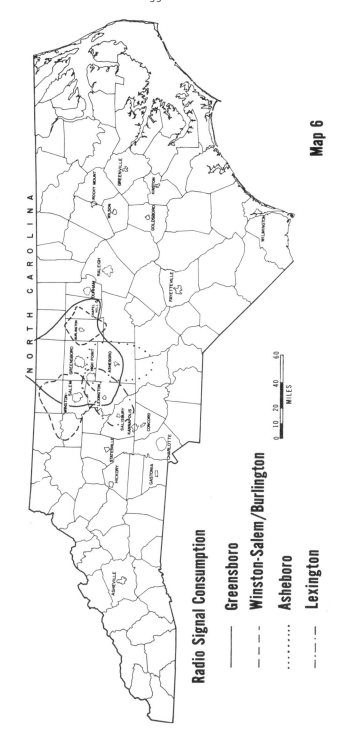

Radio Signal Consumption

——— Greensboro

– – – Winston-Salem/Burlington

·········· Asheboro

–·–·– Lexington

Map 6

MAP 7 - NEWSPAPER CONSUMPTION AREAS

The map depicts the combined 99 per cent confidence intervals
of newspaper consumption for all newspapers published in the six
major urban nodes within the study area. As is the case with the
radio signals, the newspaper consumption areas have been combined
on the map to appear as one newspaper published in each of the major
nodes, when, in actuality, each consumption area may represent more
than one newspaper. Newspaper consumption overlap is 50 per cent,
somewhat greater than radio signal consumption overlap, but considerably less than television consumption overlap. The Burlington,
Asheboro, and Lexington newspapers are consumed in the country surrounding the urban node, but the Greensboro and Winston-Salem newspapers are consumed in the entire study area. Newspaper consumption
areas are not necessarily the same as newspaper circulation areas.
The latter is derived through tabulation of papers at point-of-pur-
chase while the former is derived by asking the people of the state
which newspapers they read regularly.

35

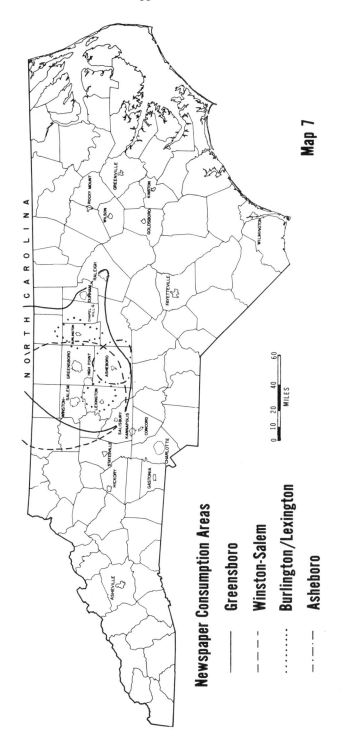

Newspaper Consumption Areas

——————— Greensboro

– – – – Winston-Salem

· · · · · · · · Burlington/Lexington

– · – · – Asheboro

Map 7

boundaries, and High Point radio captures listeners in Asheboro. The Asheboro, Burlington, and Lexington radio service areas, however, scarcely extend beyond their immediate county boundaries. Nevertheless, the forty-six per cent overlap of radio signal consumption also suggests that the Piedmont complex acts as a single market.

The consumption of daily newspapers within the area also indicates the "one-market" concept. Each node in the Piedmont complex publishes at least one newspaper, and each paper finds its way beyond the city boundaries. Newspapers published in the four smaller urban nodes tend to serve the country population beyond city boundaries as well as residents of the publishing city. The Greensboro newspapers, however, are read by some residents in all six nodes, and the Winston-Salem papers in five of the six urban nodes. The Greensboro newspapers range to the east to capture readers in the Raleigh-Durham area beyond the region. Winston-Salem newspapers, on the other hand, capture rural readers to the northwest of the Piedmont agglomeration. Newspapers of both cities are read by some residents of the Salisbury, Kannapolis, Concord agglomeration to the southwest. Map 7 shows this pattern. Newspaper consumption, like television and radio consumption, with a fifty per cent overlap suggests that the study area is a single market.

To sum up, it has been demonstrated that residents of the Piedmont urban complex often live in one urban node and shop in another. Retail interaction among the nodes of the complex is marked. Shopping patterns, in fact, suggest that the area is a single retail market rather than a group of discrete market nodes. The consumption pattern of the television, radio, and newspaper media strongly suggests that the area acts as one market.

Ginsburg suggests that residents of a dispersed city will also often live in one node and work in another.[1] Employment field delineation for the Piedmont urban complex proves this assumption correct. Workers in a dispersed city must commute daily between residence and work place just as they do in a traditional single-centered metropolitan area. In a dispersed city, however, a person may live an one urban node and work in another or live in the country and work in the city and vice-versa. Map 8 indicates that these conditions prevail in the area under study.

The employment fields depicted on Map 8 show place of work for heads-of-household who live in the individual nodes of the complex. There are places within the region where neighbors may depart for work in four different directions bound for four separate cities. The inter-nodal interaction is not as pronounced for employment fields as it is for shopping. Field overlap is only twenty-eight per cent. However, employment field delineation does indicate considerable interaction among the various urban nodes. Many local residents do, indeed, live in one node and work in another.

A labor shed is much like an employment field except that it depicts place of residence for heads-of-households who work in a particular urban node rather than place of work for heads-of-households who live in a particular node.[2] Map 9 merely supplements Map 8 since it depicts the labor shed for only one node,

---

[1] Ginsburg, op. cit. p. 8.

[2] The geographic literature on "labor sheds" and "employment fields" is voluminous. The reader is directed to James E. Vance, "Labor-Shed, Employment Field, and Dynamic Analysis in Urban Geography," Economic Geography, Vol. 36, No. 3 (July, 1960), pp. 189-220, for a landmark paper on the subject.

Greensboro, rather than for all major nodes and it depicts the labor shed for only the manufacturing segment of the labor force. However, Greensboro is a manufacturing city. Almost 40 per cent of the labor force is engaged in manufacturing goods for the local, national, or international markets, with about 80 per cent of these workers employed in textiles, tobacco, apparel, and machinery manufacturing (including electrical machinery).[1] Map 9 shows the most important segment of the Greensboro labor force and is, thus, illustrative of the distance workers travel to a manufacturing job within the Piedmont urban complex.

The Greensboro manufacturing labor shed involves four of the five urban nodes surrounding the city. Workers commute daily from Winston-Salem, Asheboro, High Point, and Burlington. The labor shed also extends toward Reidsville in Rockingham County to the north of the area under study. The Greensboro manufacturing labor shed shows that heads-of-household do live in other urban nodes or in the non-urban spaces between urban nodes and commute to and from Greensboro daily for purposes of work. It is probable that the other nodes of the urban complex are surrounded by labor sheds of a size comparable to that of Greensboro. Map 9, then, implies cross-commuting between nodes of the complex for employment purposes.

Map 10 depicts place-of-residence for people who work in the various urban nodes within the area, but the cartographic portrayal is different from that in Map 9. Map 10 utilizes the same spatially random interview sample used for shopping place as depicted on Maps 3 and 4. About twenty per cent of the sample consisted of full-time farmers, unemployed or retired people, and people who work outside the boundaries of the dispersed city. The eighty per cent

---

[1]City of Greensboro, Department of Planning, Land Use Plan (Greensboro, N.C. July, 1967) p. 43.

who journey to work are portrayed as if they worked in the centers of the various urban nodes. This, of course, is not the case, but the system utilized does make the cartographic patterns easier to understand.

Map 10 suggests a labor shed surrounding each of the major and minor urban nodes in the area. Although most of the implied labor sheds are contained within county boundaries, there are people who live in or near one of the urban nodes and work in another. Workers commute daily to Greensboro from all but Davidson County and to Winston-Salem from the counties to the east and south of Forsyth County. In fact, only Lexington and Asheboro fail to draw workers from at least one additional county.

Maps 9 and 10 do not show identical patterns. The Greensboro manufacturing labor shed depicted on Map 9 covers a larger area than does the Greensboro labor shed indicated on Map 10. This is an example of different data collection techniques yielding different results. Map 9 was compiled from nearly six hundred residence addresses furnished by fifty manufacuring plants located within Greensboro. Map 10 was compiled from only one hundred sixty residential interviews from within the five county area. (Twenty per cent of the sample of two hundred consisted of unemployed or retired people, full-time farmers, or those who worked outside the boundaries of the five-county area.) Although a representative ratio of the sample worked in Greensboro, the number of interviewees employed there was only thirty-six. A sample of thirty-six people is too small to accurately delimit a service area. Further, since no samples for Map 10 were taken outside the five-county area, no labor-shed extension beyond study-area boundaries is possible. Nevertheless, Map 10 does imply cross-commuting for purposes of employment even though it no doubt under-reports labor-shed extent.

MAP 8 - PIEDMONT URBAN AGGLOMERATION:
EMPLOYMENT FIELDS

Employment fields are areas which encompass the place of
work for residents of selected cities and towns. The areas delineated on Map 8 designate the distance that some of the workers
who live in the urban node at the center of the designated area
must travel daily for purposes of work. The map is based on the
99 per cent confidence interval of heads-of-household, so only one
per cent of these heads-of-households in the node in question
journey further than the field boundary. Obviously, the larger
the node, the smaller the employment field, since more people can
find work within the larger urban nodes. Although there are places
within the study area where neighbors may depart for work each
morning in four different directions, bound for four separate nodes,
the implied inter-nodal interaction is not as pronounced as it is
for shopping. Employment field overlap is 28 per cent. Nevertheless, some people who live in Burlington, Asheboro, High Point,
and Lexington work in Greensboro, some people who live in Lexington
work in Winston-Salem, and some people who live in Asheboro work in
High Point.

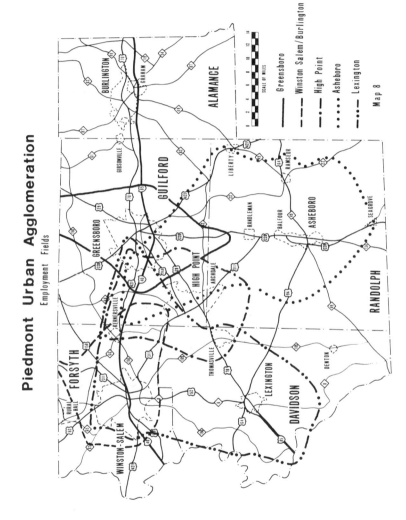

Piedmont Urban Agglomeration
Employment Fields

Map 8

SCALE OF MILES
0 2 4 6 8 10 12 14

Greensboro
Winston Salem/Burlington
High Point
Asheboro
Lexington

MAP 9 - PIEDMONT URBAN AGGLOMERATION:
GREENSBORO MANUFACTURING LABOR SHED

The map depicts that 99 per cent confidence interval of
the Greensboro manufacturing labor shed, that is, only one per
cent of the workers who work in Greensboro manufacturing plants
come from beyond the labor shed boundary depicted on the map.
Since the map shows the labor shed for only one of the urban nodes
of the study area and for only one segment of the nodal labor
force, it must be considered only suggestive of the labor sheds in
the area. Nevertheless, it is interesting to compare Map 9 with
Map 8. In the case of employment fields (Map 8) the larger the
node, the smaller the field. In the case of labor sheds (Map 9)
the larger the node, the larger the shed. In spite of this, there
are some obvious discrepancies between the two maps. Map 9, for
example, shows that some people who work in Greensboro, live in
Winston-Salem, and this is known to be so. Nevertheless, the
Winston-Salem employment field on Map 8 does not extend far enough
east to include Greensboro. This discrepancy must be due either
to sampling error or to the possibility that less than one per
cent of the Winston-Salem workers do, indeed, work in Greensboro.
Sampling discrepancies are possible since Map 8 is based on a
sample of 120 heads-of-households and Map 9 on almost 600. It is
also possible, however, that both maps are accurate and that only
one per cent of the heads-of-household do, indeed, go beyond the
employment field boundaries depicted on Map 8.

43

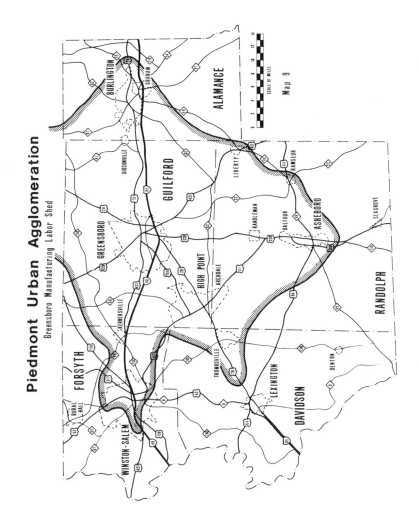

Piedmont Urban Agglomeration

Greensboro Manufacturing Labor Shed

Map 9

MAP 10 - PIEDMONT URBAN AGGLOMERATION:
LABOR SHEDS

The map suggests a labor shed surrounding each major urban
node within the study area and surrounding some of the smaller towns
within the urban region. Map 10 must be considered as only illustrative because the sample consisted of only 200 people, none of whom
live beyond the study area boundaries and forty of whom do not commute
to work. Nevertheless, Map 10 does suggest a certain amount of cross-
commuting and thus inter-nodal interaction for purposes of work,even
though it surely under-reports labor sheds for the study area. For
example, workers commute daily to Greensboro from all but Davidson
County and to Winston-Salem from the counties to the east and south
of Forsyth County.

45

Piedmont Urban Agglomeration

Labor Sheds

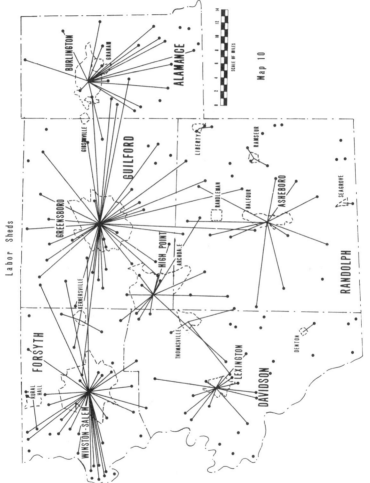

Map 10

MAP 11 - PIEDMONT URBAN AGGLOMERATION:
UNC/G COMMUTING SHED

The map depicts the boundaries of the total undergraduate

student commuting population during the spring semester of 1971.

During the period, 2,964 full-time undergraduate students commuted

to the University at Greensboro, although some may have arranged

schedules for only Monday, Wednesday, Friday or only Tuesday, Thurs-

day class attendance. The UNC/G commuting shed involves all six

of the major urban nodes within the study area and even extends

beyond the boundaries of the urban region. It is obvious from an

examination of the commuting boundary that the interstate highway

system is a great help to the student commuter. However, it is

also obvious that many students commute over national or state

highways or over secondary roads. For example, many students jour-

ney from the Eden community, located in the county directly north

of Guilford County. Eden is not linked to Greensboro by a four-lane

highway. Many students come from northwest of Forsyth County from

areas that are also not linked to Greensboro by four-lane highways.

About 75 per cent of the commuting students travel 10 miles or less,

about 15 per cent of the students travel between 10 and 20 miles,

about 5 per cent of the students travel between 20 and 30 miles,

and about 5 per cent of the students travel farther than 30 miles.

47

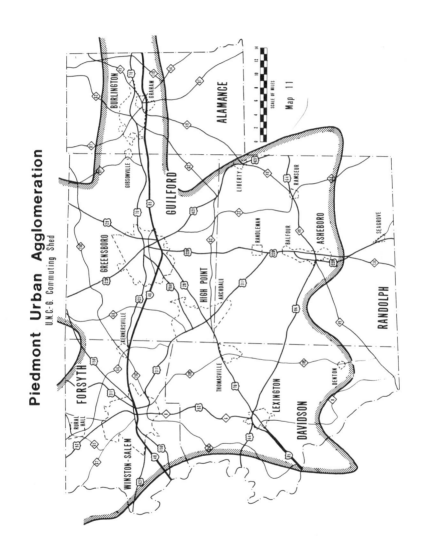

Students who commute daily to the University of North Carolina at Greensboro also constitute a labor shed of a sort. Chapter V
demonstrates that residents do perceive specialization in certain institutions of higher education, and Map 11 shows that students travel
considerable distances every day to avail themselves of the services
offered by this institution of higher learning. Map 11 is based on
the total population of the 2,964 undergraduate students who commuted
daily to the University during the spring semester of 1971. The
UNC/G commuting shed involves all six of the major urban nodes of
the Piedmont urban complex and extends well beyond boundaries of the
urban region. It is not surprising that students commute daily from
beyond dispersed city boundaries. Although branches of the University
of North Carolina are located in Chapel Hill and Raleigh to the
east and in Charlotte to the southwest, each branch specializes in
different academic offerings and thus draws students from all parts
of the state and beyond. If these students live within 50 or 60
miles of the academic institution of their choice, they often commute daily. Further, although the Piedmont complex itself contains
three major universities and several smaller colleges and technical
schools, each manages to draw students both from the urban region
and beyond.

The commanding size of the UNC/G commuting shed is explained
in part by the interstate highways traversing the urban region.
Interstates 40 and 85 connect all the urban nodes with one another
except for Asheboro, and a four-lane highway will soon connect
Asheboro with the interstate system. The interstate system traversing the urban region not only permits UNC/G students to commute
daily with relative ease but permits the cross-commuting for pur-

poses so typical of the Piedmont urban complex.[1]

Other Nodal Interaction

There are many ways in which dispersed city organization could be demonstrated beyond the retail service area delineation examined in the prior section of this chapter. Ginsburg suggests certain types of manufacturing, port functions, wholesaling, specialized entertainment and medical services in addition to the functions described earlier in this chapter.[2] A cursory look at two or three of these additional functions will help demonstrate the extent to which the area functions as a dispersed city.

Guilford County is the second largest wholesaling county in North Carolina. As of January 1, 1968, there were 810 wholesale establishments in Guilford County employing 9,100 people. These wholesalers sold 855 million dollars worth of goods. Thirty-two per cent of the wholesaling volume was distributed in Guilford County, 63 per cent to the other four counties of the area, and 5 per cent to the rest of the state and nation.[3] Guilford County wholesalers, then, do 95 per cent of their business within the five county area. These ratios suggest a considerable amount of truck travel between nodes of the study area and also reinforce the "one-market" concept.

Chapter V describes perceived specialization for recreational purposes, and it is likely that many residents do travel between

---

[1] It would be interesting but beyond the scope of this study, to go backward in time to measure the impact of the highway system on dispersed-city function. Did the construction of the interstate highway system encourage this closely spaced group of urban nodes to function as a dispersed city or was cross-commuting accomplished through other transport media before the arrival of high-speed highways? At what point in time did inter-nodal interaction become sufficient to justify the title "dispersed city," for this, or any other group of urban nodes? Only conjecture is possible at this point.

[2] Ibid. p. 6.

[3] Greensboro Chamber of Commerce, Research Division, Wholesaling in Greensboro (Greensboro, N. C. 1968)p. 2.

nodes seeking recreation. For example, it is likely that the
Greensboro Coliseum draws from within and beyond dispersed city
boundaries since it is the largest building of its kind in the state.
However, the Coliseum service area has never been measured and
delineated. There are surely many other recreational facilities
that draw residents from other nodes, but their service areas are
not known.

A cursory survey of commercial travelers staying in Greensboro motels suggests that a high percentage of these people visit
nodes of the Piedmont complex other than Greensboro, although they
stay in Greensboro at night.[1] This is another of the many potential
reasons for travel between the nodes of the Piedmont urban complex.
This interaction is, of course, necessary if the complex is to
function as an urban entity. Transportation is, as Burton and
Ginsburg suggest, the crux of dispersed city organization and must
be given careful consideration if dispersed cities are to function
efficiently. Chapter IV reports the transportation planning efforts of the local governmental council.

Summary

The evidence presented in this chapter demonstrates that the
Piedmont urban complex functions as a single urban unit in certain
respects. Downtown-trade-area overlaps strongly imply cross-commuting among urban nodes for shopping purposes. Interviews with
residents of the urban region reinforce this assumption in that the
entire sample shopped in at least one central business district
other than the closest and shopped there as often as in the closest.
Further, two-thirds of the sample offered a third central business
district shopping choice, and one-tenth of the sample offered a

[1]Charles R. Hayes, Greensboro Recreational Survey, Greensboro, North Carolina, Chamber of Commerce 1972, pp. 1-10.

fourth choice. Third and fourth shopping choices were visited only
infrequently, however, suggesting limits to the extent that interaction takes place. Central business district shopping patterns suggest
that the complex functions as a single marketing area, albeit an
open system subject to exogenous influences and activities.

Work service areas also indicate that residents often live in
one node and work in another. Moreover, the patterns of communications
media consumption suggest a single market; commuting students engage
in considerable travel within the complex; and wholesale distribution
suggests considerable inter-nodal truck travel.

The concept of a group of closely spaced, but discrete urban
nodes functioning economically as a single urban unit-- a metropolis
without a core-- fits the Piedmont urban complex reasonably well,
albeit imperfectly. On the basis of the evidence thus far presented
the area under study would seem to be an incipient or partial dispersed city.[1] Perhaps the Piedmont urban region is still developing dispersed city functions. Nevertheless, for lack of a better
example in the geographic literature, and although the term is not
a perfect fit, it can be tentatively applied to the area under study.

---

[1]This proposition is based upon the question of interaction
which has been dealt with at least in part, but the corollary
question of areal specialization remains open. Do some residents
live in Winston-Salem and shop in Burlington, or live in Lexington
and work in High Point, or live in Asheboro and read a Greensboro
newspaper because of inter-nodal specialization in high-level
goods and services? It is doubtful. However, some light can be
shed on these questions if it can be discerned how residents perceive such specialization. Resident perceptions are examined in
Chapter V.

Chapter III

THE DEVELOPMENT OF THE AREA

"The south is dissimilar to much of the rest of the
nation in its settlement pattern and in many of its social
and economic characteristics; all are deeply rooted in the
region's history. It is still a region of largely medium
and small-sized towns, rather than large urban centers. It
follows, therefore, that much of the South's manufacturing
is associated with medium- and small-size towns. The bulk
of the South's industry is located in counties whose largest towns have a population of less than 100,000 and com-
paratively little is found in major industrial areas."[1]

Introduction

The Lonsdale and Browning description of the settlement struc-

ture of the South quoted above well applies to the Piedmont urban

complex segment of the "South". Manufacturing is dispersed through-

out the region in the smaller towns and in rural areas as well as

in the major urban nodes. Since much of the economy of the Piedmont

complex depends on manufacturing, and since all cities are, of course,

products of their past, a consideration of the development of the

Piedmont complex over time can help explain the present morphology

---

[1]Richard E. Lonsdale and Clyde E. Browning, "Rural-Urban
Locational Preferences of Southern Manufacturers," Annals of the
Association of American Geographers, Vol. 61, No. 2 (June 1971),
pp. 255-268.

of the urban region.

The familiar story of the impact of New England textile manufacturing firms on the agrarian economy of the South is, at least for this area, oversimplified to the point of error. The story goes that around the turn of the last century the cotton textile industry of New England commenced its southward migration attracted by cheap trainable labor pushed from the farms of the southern Piedmont by mechanization. The industry had been pushed from the north by antiquated plant facilities and a difficult labor situation.[1] In fact, however, the Piedmont area had several textile manufacturing plants by 1840. It is difficult to say just how many textile manufacturing plants opened and closed between 1830 and 1850 because no one really counted them, but ten plants is a conservative figure for the antebellum industrial boom.[2] Many of these plants re-opened shortly after the Civil War, a good ten to twenty years before the great "textile migration from the North." The Piedmont, then, has had a history of textile (and other) manufacturing well before 1900 and could capitalize on the labor union strife and antiquated facilities in New England and compete with it with expertise, experience, and price in the competition for textile markets.

---

[1] See, for example: a) Victor R. Fuchs, Changes in the Location of Manufacturing in the United States Since 1929 (New Haven, Yale University Press, 1962) pp. 87-258 -or- b)Calvin B. Hoover and B.U. Ratchford, Economic Resources and Policies of the South (New York, Macmillan Co., 1951)pp. 33-116, -or- for that matter almost any economic geography textbook. It should be noted that not all researchers agree that the South grew industrially at the expense of the North. See for example a) Glenn E. McLaughlin and Stefan Robock, Why Industry Moves South (Kingsport, Tenn., Kingsport Press, 1949)pp. 3-75 -or- b) National Planning Association, Committee on the South, New Industry Comes to the South, Washington, D.C. (1949) pp. 1-11.

[2] See, for example, Bill Sharp, A New Geography of North Carolina (Raleigh, N.C. Sharp Publishing Company, 1958) pp. 946-1040, or Ethel S. Arnett, Greensboro, North Carolina: The County Seat of Guilford (Chapel Hill, N.C. University of North Carolina Press, 1955) pp. 1-10.

The antebellum textile industry of the area manufactured its products for local and regional markets and, by extension, for those of the South in general, but not for those of the North, for the railroad did not reach the area until after 1850, and even the "plank road" did not arrive until about 1840. Full wagons generally traveled south because the land sloped in that direction.When rails were laid in the area and after the Civil War was terminated, transportation ties with the North were developed, and, for the first time, markets outside the South became accessible to the products of the region.

The railroad provided transportation northward, and price differentials opened the northern markets to textile and other manufacturers of the Piedmont. Textiles could be manufactured at less cost in the South because labor was cheap and plant facilities modern. The textile boom during the period 1890 to 1925 did not result from migration from New England or for that matter from anywhere else. Every major textile manufacturing plant in the area during the period was started with local capital, local management, and local entrepreneurship.

The following paragraphs trace the development of the study area on a county-by-county basis. In each case, settlement, livelihood, and the introduction of manufacturing is considered.

Forsyth County

Forsyth County, the northwestern-most of the five counties in the study area, was first settled by the Unitas Fratum, a religious group which had been persecuted in Bavaria, had fled to Moravia, and thus had come to be called Moravians, and which was, in 1749, granted by the British Parliament the right to practice their religion freely anywhere within Great Britain and its colonies.[1] The Parliamentary testimony of William Penn that the Moravians were sober, reliable,

---

[1] Levin T. Reichtel, The Moravians in North Carolina (Baltimore, Genealogical Publishing Co., 1968), p. 18.

religious, and successful brought many offers to the group by proprietors seeking such people as colonists. The Moravians accepted
the offer of Lord Granville (the last of the original proprietors
of North Carolina) of 198,965 acres in the North Carolina Piedmont
where Winston-Salem now stands. According to Reichtel, the decision
was made because of the size of the tract not the quality of the land,
for the Moravians were craftsmen not farmers.[1]

Construction of a planned "new town", Salem, was begun in
1766 as a place designed for the manufacture and sale of cloth,
leather, food, construction materials, stoves, and smoking pipes,
with other manufactured items to be added as demand warranted.[2]
Salem quickly became the most important place in the region because
of the range of goods and services offered. As such, Salem was the
logical site for the county seat when the North Carolina legislature
carved the state into county administrative districts. All North
Carolina counties were of a size to permit round-trip travel of not
more than one day from county boundary to county seat. Thus, the
range of administrative authority included in the landscape is a
function of the transportation technology of well over a century and
a half ago. The practice of sub-dividing the existing but very large
North Carolina counties into smaller units and designating a county
seat as near the geometric center of each unit as topography permitted
was a one-at-a-time political decision in the State legislature.
Nevertheless, this decision was made for Forsyth County[3] and, as will
be noted, for the other four counties of the Piedmont urban complex.

---

[1] Ibid. pp. 18-19.

[2] Edward Roundthaler, The Memorabilia of Fifty Years 1877-
1927 (Raleigh, N.C. Edwards & Broughton Publishing Co., 1928),
p. 48 and Robert W. Neilson, History of Government (Winston-Salem,
City of Winston-Salem, 1966), pp. 15-16.

[3] Ibid. pp. 15-16.

Distance friction was the key element in the decisions, and round-trip travel of one day or less by horse and wagon from county boundary to county seat at the center of each county limited the size of Forsyth and the other four counties within the area. This practice also determined the spacing of the major urban nodes because each county was required to build or designate a county seat near the center of the county.

Salem declined the honor of the county administration center but sold the County Commissioners enough land for the purpose just north of but contiguous to Salem. Thus, the "new town" of Winston was built for administrative purposes with a courthouse square one mile north of Salem Square. Although it was more than half a century before the two joined administratively to become Winston-Salem, according to Roundthaler, it was obvious from the first that they would function as a single urban unit.[1]

Unlike certain other nodes in the study area, Winston-Salem experienced no real population or manufacturing growth until after the Civil War. In 1879 the combined population of the two nodes was 1,348. The railroad reached the area in 1873, and by 1880 the population had increased to 4,194.[2] The rapid population increase is attributable to the establishment, during the period, of the R.J.Reynolds tobacco processing factory, a local industry development, but one with national markets.[3] The Reynolds' factory brought the first wave of population growth to Winston-Salem and marked its entrance into the industrial age. The manufacturing base of Winston-Salem today is quite diversified and generates more than 40 per cent of the total jobs in the city. Even though most of this manufac-

---

[1]Roundthaler, op. cit. pp. 57, 79, and 103.

[2]Ibid. pp. 103-150.

[3]Hammer, Green, Siler, Associates, Forsyth County's Economic Prospect (Washington, D.C., April, 1970) p. 7.

turing was added after World War II, the manufacturing diversity that
characterized early Salem also characterizes modern Winston-Salem.

Winston-Salem today reflects the early settlement of Salem
by a group of people who deliberately established a manufacturing
city rather than a rural trading center. County administration
followed with the founding of Winston, and early crafts developed
into manufacturing after the Civil War. The cigarette, apparel, and
electrical machinery industries provide most of the factory jobs
today in Winston-Salem, and the two periods of rapid population
growth (post-Civil War and post-World War II) can be attributed
directly to industrial expansion[1] Nevertheless, until the establishment of a Western Electric plant in 1946, 95 per cent of the manufacturing industry was a product of local capital, local ownership,
and local management.[2]

Guilford County

Guilford County, the county to the east of Forsyth County was,
like Forsyth, settled, at least in part, by craftsmen. English
Quakers interested in worship and education and who proposed to support these endeavors through the manufacture and sale of handicrafts,
settled the western portion of what is now Guilford County.[3] Although
the central and eastern portions of the county were settled by Scotch-
Irish and Germans interested in agriculture, the Quaker craftsmen
have had a lasting impact on the local way-of-life.

Furniture, clothing, iron plows, and guns were among items
manufactured by the English Quakers in the early 1830's. In fact,

---

[1]Ibid. p. 5.

[2]Ibid. p. 5 -and- Winston-Salem Chamber of Commerce,
A Half Century of Progress 1855-1935 (Winston-Salem, Sept. 1935)
p. 20.

[3]C.A. Paul, Greensboro Daily News (Greensboro, N.C. May 29,
1971) p. C3.

the very first steam cotton mill in North Carolina began producing fabric during the same period. The mill was located in Greensboro.[1] A decade later there were eighteen manufacturing firms in Greensboro. Textiles, clothing, lumber, and metal specialities were among items manufactured for local consumption and for export to neighboring states.[2]

Guilford officially became a county in 1771. Guilford County, like Forsyth County, was created of a size to permit one-day round trip travel by horse and wagon from periphery to center, and a county seat was planned for the geometric center of the county. Although the county seat was temporarily located off-center, the permanent county seat, Greensboro, was created in 1809 as near the center of the county as terrain permitted.[3] Greensboro was a "new town" created for administrative purposes but destined for a manufacturing role.

The Greensboro manufacturing thrust in the 1830's noted in prior paragraphs, was followed by an even greater thrust in the mid-1850's with the arrival of the North Carolina railroad. The rails provided transportation to formerly inaccessible markets and provided transportation from these markets for in-migrants to work in the newly established factories. In 1829 the population of Greensboro was 369, but by 1859 had increased to 1,400.[4] The Civil War wiped out manufacturing in Greensboro, but many of the factories

---

[1]Ethel S. Arnett, Greensboro, North Carolina: The County Seat of Guilford (Chapel Hill, N.C., University of North Carolina Press, 1955), pp. 1-10.

[2]Ibid. pp. 1-10.

[3]Sallie W. Stockton, The History of Guilford County, North Carolina (Nashville, Tenn., Gaut-Ogden Co., 1902), pp. 1-125.

[4]Robert E. Register, Greensboro Daily News (Greensboro, N.C. May 27, 1971), p. C38.

had reopened by 1875[1] with the textile boom more than a decade in the future.

The textile boom became a reality in Greensboro between 1890 and 1925, but the period also brought many other types of manufacturing. Textile machinery followed textiles to the area within a few years and soon spun off into a diversified machinery manufacturing complex.[2] Clothing and wood products were also early arrivals on the Greensboro Manufacturing scene.[3]

The manufacturing growth of this period is associated with a startling population growth. In 1900 Greensboro counted 10,035 persons; by 1930 almost 60,000 people called the city home. Guilford County increased from almost 40,000 to more than 133,000 people during the same period.[4] Manufacturing brought people to work the factories, but every major company that came to Greensboro during this boom period was result of local capital, local management, and local expertise.

High Point, also within Guilford County, got a later start than did Greensboro. When the North Carolina Railroad survey party reached the crossing of the plank road connecting Salem to Fayetteville (North Carolina) they pronounced it the "high point" on the proposed rail line. At that time, in 1852, the "high point" was a wilderness.[5] The transport situation offered by the junction of the railroad and the plank road assured the growth of the new community. The first train arrived in 1855. By 1857, 159 votes

[1]Stockton, op. cit., pp. 1-25.

[2]Charles R. Hayes and Norman W. Schul, "Why Do Manufacturers Locate in the Southern Piedmont?" Land Economics, Vol. XLIV, No. 1 (Feb. 1968), pp. 117-121.

[3]Ibid. pp. 117-121.

[4]Charles O. Forbis and D. Parker Lynch, Population: Guilford County, N.C. (Greensboro, N.C., Guilford County Planning Department, 1965), pp. 7-8.

[5] Robert Marks, Greensboro Daily News (Greensboro, N.C. May 29, 1971), p. D5.

were cast in the first general election, and in 1859 the town was chartered.[1] The first factory in High Point was established just prior to the Civil War. The plant manufactured wood products, a harbinger of the furniture industry to come.[2] Nevertheless, it would be another quarter century before the furniture industry would arrive.

After North/South hostilities had ceased, the High Point transport situation made it a natural trans-shipment point for lumber. Wagon loads of lumber traveled the plank road to the rail terminal for transshipment. A group of local business men noted the cost advantages of the break-of-bulk situation and organized the first furniture manufacturing plant in 1888.[3] By 1900 there were 33 furniture plants in the city; by 1914, 107.[4] In 1904, in order to tap the labor potential of the female members of the furniture workers' families, a hosiery mill was established. Thus a symbiotic relationship was developed between the two activities. The manufacture of hosiery is High Point's leading industry today.[5]

Manufacturing had an early start in Guilford County, and the fact that many of the early settlers were craftsmen had much to do with the early start. The coming of the rails opened new markets for existing and new manufacturing industry, and industry brought people to work in the factories. The War between the States wiped out the early industry in the county, but many plants resumed production shortly thereafter. The textile boom at the turn of the century brought industry of many types of the country. Today,

[1] Ibid. p. D5.

[2] Ibid. p. D5.

[3] Ibid. p. E50.

[4] Ibid. p. E50.

[5] Hammer, Green, Siler Associates, The Guilford County Economy (Washington, D.C., 1966) p. 102.

textiles, furniture, apparel, and machinery account for nearly 70
per cent of the manufacturing.[1]

Alamance County

Alamance County, just east of Guilford County, like Forsyth and Guilford Counties, was settled, in part,by craftsmen.
Between 1700 and 1755 Germans from the Rhine area and English
Quakers from Pennsylvania organized several small communities.
Whereas the German settlers were interested in agricultural activities, the Quakers came to build towns for educational purposes
and for the manufacture and sale of craft items.[2] Nevertheless,
it was to be nearly a century before manufacturing arrived. In
1837, a cotton textile mill was built on the banks of the Great
Alamance Creek, the fourth such mill in the state. Fifteen years
later the mill produced the first colored cloth in the South.[3]

Alamance County was formed by the state in 1849, in
order to create an administrative area small enough so a trip from
boundary to center and back, by horse and wagon, would take no more
than one day. The county seat was to be established as close to
the geometric center of the county as the terrain permitted.[4] Distance friction dictated the county size, and Graham, the county seat,
was laid out as close to the center as terrain permitted. It was
the railroad, however, that provided the first real industrial thrust
for the county. When the North Carolina railroad arrived in the mid-
1850's, the citizens of Graham objected to the plan for locating

[1] Fautus Company, Industrial Location Appraisal: Guilford County (New York 1967) p. 3.

[2] George Bucher, Investigation of Local Resources for the Social Studies in Alamance County, A Looseleaf Collection of Source Materials for Local History and Social Problems (Burlington, N.C. 1939)p. 1.

[3] Walter Whitaker, "E.M.Holt and the Cotton Mill", Centennial History of Alamance County (Burlington, N.C. Chamber of Commerce, 1949), pp. 96-103.

[4] Donald E. Bolden, Alamance Battleground, Bi-Centennial, Commemorative Souvenir Program (Burlington, N.C. May 1971) p. 28.

maintenance facilities within the town; consequently, the rail line
passed north of Graham, creating the adjoining town of "Company
Shops" (later Burlington).[1] By 1860, there were 5 cotton textile
mills, 4 tobacco factories, and several other manufacturers located
in the Burlington/Graham urban node.[2]

The Civil War erased manufacturing industry from Alamance
County. It was 1881 before manufacturing returned to the County.[3]
Although the first manufacturing plant organized after the Civil
War produced sashes and doors and later furniture, by 1879 there
were six cotton textile mills in the county, and by 1890, seventeen.[4]
Population growth paralleled industrial growth. In 1880 Burlington
had 817 residents and by 1890, 1,716.[5]

The cotton textile industry is the foundation of the Burling-
ton/Graham -Alamance County economic base today, as it was a century
ago. However, there was no sharp impact at the turn of the century
such as Greensboro and Guilford County experienced. Rather, the
industry grew over a period of years with the addition of individual
small plants, each the product of local capital, labor, and entrepreneurship, until by 1924 twenty-two cotton textile mills operated
within the county.[6] The year 1923 marks the real textile impact on
the area. In that year a group of local businessmen financed a new
textile mill. This textile manufacturer was to become the world's

[1]Ibid. pp. 19-25.

[2]Bucher, op. cit., p. 5.

[3]Mary L. Macintosh, "History of Elon College", Diamond
Jubilee, 1893-1968 (Elon College, N.C. 1968), p. 1.

[4]Whitaker, op. cit., pp. 163-165.

[5]U.S.Dept. of Commerce, Bureau of the Census, Census of
Population (1890).

[6]Whitaker, op. cit. p. 165.

largest.[1] The population growth of the area reflects the success
story and consequent growth of one textile manufacturer. The population of Burlington-Graham increased from 8,318 in 1920 to 16,597
in 1940 as Burlington Mills expanded operations. The decade of the
1940's, though, marked the greatest expansion of Bur-Mils and the
establishment of a Western Electric factory, the first major industrial migrant from the north. Burlington/Graham reflected these
events with a population increase of almost 13,000 people; from
16,957 to 29,588.[2]

Textile manufacturing (weaving and knitting)currently
acconts for almost 73 per cent of the manufacturing jobs in Alamance
County. Twenty-five per cent of the manufacturing jobs are about
equally distributed among electrical machinery, furniture, and food
processing. The remaining two per cent are scattered among several
other diversified manufactural industries.[3]

Alamance County, centered on Burlington/Graham,is, like
Forsyth and Guilford counties, a product of its past. Time required
for the round trip from periphery to center limited county size and
led to establishment of the principal urban node. Manufacturing
industry followed crafts to the county, each a product of local
effort. The proclivity of southern manufacturers to locate in small
towns and throughout the countryside permitted some industrial location there rather than a few miles to the west in older, larger, and
better established Greensboro and Guilford County. The anti-locali-
zation trend for southern manufacturers noted by Longsdale and Browning is a century old, at least in the Piedmont urban complex.

[1] Donald E. Bolden, "The History of Burlington Mills",
unpublished paper (Chapel Hill, The University of North Carolina,
1955), pp. 1-10.

[2] Hammer, Green, Siler Associates, Downtown Burlington: An
Analysis of Its Economic Potential (Washington, D. C. July 1967),
pp. 23-25.

[3] Ibid., pp. 23-25.

64

Lonsdale and Browning demonstrate that manufacturing firms
throughout the southeastern United States locate in small cities and
towns and even in the countryside as well as in or near the principal
urban node of a region. Nevertheless, the preference is for the
smaller places rather than the major urban centers, and this characteristic is intensifying. Lonsdale and Browning suggest that the
character of southern manufacturing -- labor intensive, with a low
profit margin, and labor economies essential to maintain a competi-
tive market position -- tends to discourage manufacturing competition
in larger cities with tighter markets and higher wages.[1] The fact
that manufacturing industry located in Burlington/Graham and throughout Alamance County rather than in older and larger Greensboro a
few miles to the west, supports the Lonsdale-Browning observation
and suggests that the trend is at least a century old. Had this
trend not been in effect in the study area during its earlier period
of development, it is likely that one of the urban nodes would
have become dominant, since population growth is now, and has long
been, tied closely to manufacturing growth throughout the southeastern states;[2] and a dominant city surrounded by satellites is
not a dispersed city by definition.

Randolph County

English Quakers, Germans, and Scotch-Irish settled Randolph
County in the early eighteenth century. Just as in Guilford and
Alamance Counties, the Quakers were craftsmen and the Germans and

[1]Lonsdale and Browning, op. cit., pp. 255-256.

[2]See for example a) Frank T. De Vyver, et al. Labor in the
Industrial South (Charlottesville, Va., Michie Co., 1930) p. 4-13;
b) Victor R. Fuchs, Changes in the Location of Manufacturing in the
United States Since 1929 (New Haven, Yale University Press, 1962)
pp. 87-259; c) Calvin B. Hoover and B.U.Ratchford, Economic Resources and Policies of the South (New York, Macmillan 1951)pp.34-144;
d) Curtis H. Braschler, "Importance of Manufacturing in Area Economic Growth," Land Economics, Vol. XLVII (Feb. 1971) pp. 109-111;
e) Truman Hartshorne, "The Spatial Structure of Socioeconomic
Development in the Southeast," Geographical Review, Vol. 61, No. 2
(April, 1971) pp. 46-51.

Scotch-Irish farmers.[1] As was the case in Guilford and Alamance Counties, the Quakers of Randolph County quickly established schools and churches and began producing and selling craft items in order to support their endeavors.[2] Like Guilford and Alamance Counties, these craft operations developed into early industry.

Randolph became an official county in 1779. The county was formed from part of Guilford County to the north, so that residents need not travel so far to conduct county business. Although the first county seat was off-center, in 1793 the seat of county government was moved as close to the geometric center of the county as the terrain permitted.[3] At that time, the site of the county seat, Asheboro, was only a crossing of two trails and contained no structures. In fact, Asheboro functioned as little more than county administrative center for the next hundred years. By 1850, Asheboro could boast of only 103 people and by the year 1900, less than a thousand.[4]

The rugged terrain of Randolph County kept the area isolated until late in the nineteenth century. The north Carolina Railroad bypassed the county initially, and it was not until 1896 that a rail spur was built into the county from High Point in order to haul lumber to the expanding furniture industry. The rugged terrain, however, provided water power potential for the early industry of the county.[5] In 1836 the first cotton textile mill was established in

---

[1]Fred Burgess, Randolph County: Economic or Social (A Laboratory Study of the University of North Carolina) (Chapel Hill, Department of Rural Social Economics, 1924), p. 9.

[2]Ibid., pp. 9-10.

[3]Ibid., p. 12.

[4]United States Bureau of the Census, Statistical Atlas of the United States (Washington, D.C., Government Printing Office, 1914).

[5]Bill Sharp, A New Geography of North Carolina (Raleigh, N.C., Sharp Publishing Company, 1958), p. 1041.

the county on the Deep River. By 1850, there were four cotton mills
within the county, each utilizing the fall of the rivers for power.[1]

Although Randolph County's first industry was cotton textiles,
the most important industry of the decade of the 1840's was lumbering.
A plank road, connecting Fayetteville with Salem, built in 1839,
passed through the county. Wagon loads of lumber traveled the road,
first for construction use, then for rail cross ties, and finally,
for use in the furniture industry in High Point. Lumbering is still
an important industry in the county.[2]

The inaccessibility that plagued Randolph County in its early
days was a boon during the Civil War period. The War between the
States virtually bypassed the county. Nevertheless, it was not until
the railroad provided access to the north at the turn of the century
that Randolph experienced an important impact from manufacturing
industry. The years between the Civil War and the year 1900 showed
very little industrial or population growth.

The textile boom at the turn of the last century had great
impact on Randolph County, as it did on the counties to the North.
The railroad not only opened Northern markets to the area but
brought in coal to power the industry. The use of fossil fuels
quickly replaced water power for industry within the country.[3]

Population growth paralleled industrial growth. In 1900
Asheboro was the residence of 992 people; by 1930, 5,021 persons
called the city home. Randolph County contained 28,232 people in
the year 1900. By 1930, 36,259 persons lived within county boundaries. In 1894 there were 14 textile mills within the county;
by 1925, more than one hundred.[4]

[1]Burgess, op. cit., p. 21.

[2]Ibid., p. 27.

[3]Ibid., p. 29.

[4] Sharp, op. cit., p. 1029.

The cotton textile industry is Randolph County's largest
employer today; wood and wood products the second largest.[1]
Although most of the industry within the county is related to
these two industries, industrial in-migration has recently shown
diversification. Clothing, shoes, plastics, and paper packaging
are recent additions to the county industrial scene.[2]

Randolph County must certainly be considered one of the area's
industrial components. Nevertheless, the county retains a rural
flavor even today. Much of the county is still forested and more
than half the county's residents reside outside city bounaaries.
More than half of these rural residents, however, are part-time
farmers and part-time factory workers.[3] Although Randolph County
is an industrial partner in the southeast Piedmont, many of her
residents prefer the rural life. The part-time farmer and part-
time factory worker is not unique to Randolph County, but this way
of life is more prevalent here than in the counties to the north.

Davidson County

The familiar story of the state legislature carving a county
from a larger area and designating a seat of government near the
geometric center is repeated for Davidson County. The county was
formed in the early 1820's in order to facilitate access of citizens to a county seat. Lexington was designated the county seat.
Unlike the other county seats in the area, however, Lexington at
the time was already an existing town.[4] The small town that was

[1] Asheboro Chamber of Commerce, Asheboro, North Carolina,
Basic Industrial Data (Asheboro, N.C. undated) p. 1.

[2] Ibid., pp. 1-2.

[3] Charles R. Hayes, D. Gordon Bennett and Linda Sowers,
Phase I Regional Development Guide (Greensboro, N.C. Piedmont
Triad Council of Governments, 1971), pp. 1-10.

[4] Lexington Chamber of Commerce, Here's Lexington (Lexington,
N.C., undated brochure) pp. 1-2.

68

Lexington had developed as a station on the Greensboro/Salisbury
road. The people of the area needed a shipping point for agricultural surpluses and craft items, and Lexington filled this need.[1]

Davidson County was settled primarily by people of German
extraction. These people considered themselves farmers rather than
craftsmen. Nevertheless, events forced these agriculturally oriented
people to turn to crafts for supplemental income. Agriculture in
Davidson County prior to the Civil War was largely subsistence.[2]
As the moderate surpluses did not fulfill the residents' needs for
financial income, cottage industry was born. Craft items and certain agricultural products were transported to Lexington for ship-
ment and sale.[3]

Davidson County today retains much of the flavor of the
early nineteenth century. Many heads-of-household prefer rural to
urban living but commute daily to factory jobs in the urban nodes.
Part-time farmer/factory workers are numerous in Davidson County,
and many small farms whose owners farm only part time occupy the
countryside.[4]

The German farmers of Davidson County, then, became craftsmen
in spite of themselves and provided the nucleus of the labor force
when the first textile boom hit the county in the 1830's. The first
textile manufacturing plant in the county, a product of local initiative, was established in Lexington in 1838. By 1860, three years
after the railroad crossed the county, there were 102 factories in
Davidson County. The processing of cotton and hemp into cloth and

[1]City of Lexington, Land Development Plan (Lexington, N.C.
1961) pp.1-18.

[2]Ibid, pp. 1-18.

[3]Ibid., pp. 1-18.

[4]Hayes, Bennett, and Sowers, op. cit., pp. 1-10.

the manufacture of wood products including furniture were the main
industries, then, and, with the exception of hemp, remain the main
industries today.[1]

As was true in most of the other counties of the area under
study, the War Between the States had a deleterious effect on the
industry of the county, and it was not until the decade of the 1880's
that industry returned. The textile boom started in the mid-eighteen
eighties, and the industry was followed closely by a furniture industry in migration.[2] Population growth in Lexington paralleled industrial growth. In 1890 there were 611 people living in the county
seat. By 1930, almost 10,000 people called Lexington home.[3]

The urban nodes of Davidson County rest on a manufacturing
base of textiles, clothing, and furniture. However, over the last
two decades industrial diversification has taken place largely as a
result of industrial migration from Forsyth and Guilford counties.[4]
This again demonstrates the proclivity of industry to locate in small
towns and rural areas as well as in larger urban places.

Dispersed City Morphology

Chapter II has explored the extent to which the group of
nodes comprising the Piedmont urban complex operate as a functional
unit and, therefore, may be defined as a "dispersed city". This chapter has traced briefly the development of the area. It is now possible to generalize as to the "why" of dispersed-city development.
Why did this area, and certain other urban regions in the world,
develop as polynucleated rather than mononucleated urban centers?

---

[1] Lexington Chamber of Commerce, Lexington, North Carolina
(Lexington, N. C. 1971) pp. 1-10.

[2] Lexington Chamber of Commerce, Lexington, North Carolina,
op. cit., pp. 1-10.

[3] U.S. Bureau of the Census, Census of Population (1890-1930).

[4] Hayes, Bennett, and Sowers, op. cit., pp. 1-10.

Burton suggests that the emergence of dispersed cities depends in part upon the level of transport technology operating in the formative stage of the settlement pattern.[1] Transport technology operating in the formative stage of the settlement pattern is certainly step one in dispersed city formation. In the Piedmont urban complex the urban nodes were closely spaced so citizens could avail themselves of the administrative services of county seat without excessive travel difficulty by the then transport medium of horse and wagon. The spacing of all but one of the major urban nodes was set by transport technology at the formative development stage as determined by administrative decisions. There are surely other dispersed cities in which nodal spacing was determined for the same reasons. The distribution of urban nodes in yet other dispersed cities may have been determined by the desire to supply other services - commercial, religious, educational, et al. - to a surrounding population equally limited by primitive transport technology. In any case, transport technology at the formative stage of development dictated that the urban nodes of dispersed cities would be closely spaced. (of all class)

Transport technology is not, however, the whole story. The new labor intensive low-profit-margin manufacturing plants were, as Lonsdale and Browning have pointed out, distributed in the smaller cities and towns and in the countryside, seeking "looser" labor markets and lower wages. This trend precluded the dominance of one node since population growth in the southern Piedmont was and is closely tied to manufacturing growth. No one node, therefore, could become large enough to dominate the urban complex. Since local people dictated manufacturing location, when a plant was built in Greensboro, the next was built in Winston-Salem or Burling-

---

[1] Burton, "A Restatement" etc., op. cit., p. 286.

ton, or High Point, or somewhere in between. This manufacturing locational rotation accounts in part for the present morphology of the Piedmont urban complex. Nevertheless, the very fact of manufacturing locational rotation is strange. Classical location theory argues that agglomeration, _ceteris paribus_, means economies, whereas dispersion means higher costs. Does the labor, rather than a transport orientation of Southern industry permit economical dispersion as Lonsdale and Browning suggest, or are costs really higher, and people prefer it that way?

Transport technology and manufacturing locational rotation at the formative development stage have greatly influenced the multinuclear morphology of the Piedmont urban complex. An additional reason for this sort of morphological development in North Carolina is the extreme importance of the county seat to "Tar Heels." Senator Sam Ervin speaks of "coat week" (court week) as the most important period of the year in past decades to North Carolineans.[1] During this quarterly fortnight county residents journeyed daily to the county seat to attend the trials and to be entertained by medicine shows, carnivals, and by local urban amusements. William Trotter points to the importance of the North Carolina county seats today as the sites of local stock car races.[2] County residents spend the day at the dirt track to watch the young drivers in their "souped-up" stock cars compete for cash prizes. These residents remain after the race to eat and drink. This local support of the county seat tended to keep each a viable growing city. Even today a "good" trial, a stock car race, or an important high school basketball game will bring rural residents to town for the day.

---

[1]Thad Stem, Jr. and Alan Butler, Senator Sam Ervin's Best Stories (Durham, N.C., Moor Publishing Company, 1973), pp. 9, 10, 11-16, and 17.

[2]William Trotter, "Four Hour Thunder", Red Clay Reader, No.6 (Charlotte, N.C. Southern Review, 1969), pp. 87-99.

It is possible that other dispersed cities of the world may
have maintained relatively even nodal growth rates in the same way
or through other means. For example, the location of commercially
extractable mineral resources as in southern Illinois, might have
kept closely spaced urban nodes growing at about the same rate until
the resources became exhausted. At this point each of the closely
spaced nodes would need to offer some special service in order to
maintain the same rate of growth as its neighbors. Still other
closely spaced urban nodes might have maintained similar growth
rates because each learned early to specialize in some high order
good or service. At any rate, relatively even growth rates over
the past decades is the other necessary dispersed city ingredient.
The maintenance of the dispersed city structure depends, of course,
on inter-nodal specialization and inter-nodal communication and
transportation. The possibilities of maintaining that structure
for the future are examined in Chapter IV.

Chapter IV

POPULATION AND PLANNING IN THE AREA[1]

"Number of people is probably the most basic information
about the earth and its regions"[2]

General Population Characteristics

Growth of population over the past thirty years is perhaps

the outstanding human event recorded for the Piedmont urban complex.

In 1970 almost 800,000 people called the multi-nodal urban region

home. In 1940 only a few more than half that number lived in the

area. Population growth within the area over the past three decades

reflects national growth trends. Population growth has a particular

impact on transportation here because of extreme dependence on the

private automobile and the considerable amount of inter-nodal travel.

It is a reasonable assumption that with more people living

in the area, more people will travel within the area, and thus

travel difficulties will increase for all. On the other hand, it

could be hypothesized that during the thirty years from 1940 to 1970,

---

[1]Data for this chapter are from the following sources:
1) United States Bureau of the Census, Census of Population (1970)
-and- Charles R. Hayes and D. Gordon Bennett, "Population and Planning in the Piedmont Dispersed City, " paper read at the N.C. Population Center, Chapel Hill, N. C. (Feb. 1973).

[2]Glenn T. Trewartha, A Geography of Population: World Patterns
(New York: John Wiley & Sons, 1969), p. 3.

travel difficulties actually were reduced. Personal incomes rose substantially in the United States over the past three decades. Automobile transport technology improved considerably over the same period. Surely, as personal income rises, that portion assigned to transportation becomes a smaller share of total income. It is also reasonable to suggest that as transportation is improved, the time devoted to personal transportation decreases. If these assumptions are correct, the problem of overcoming the friction of distance diminishes. If travel difficulty was reduced in the area from 1940 to 1970, surely inter-nodal interaction increased and an increase in inter-nodal interaction enhances its "dispersed-city" character. Still, an automobile-centered transport system can become overloaded in any urban area. Whether distance friction in the area was increased or decreased over the past thirty years, inter-nodal transportation has now reached the critical stage as evidenced by the criticisms of the area residents (Chapter V).

The population growth rate in the five-county area declined in the decade of the sixties from previous decades just as it did in other cities of the country. Between 1960 and 1970 about 20,000 people were added to the urban region through net in-migration, and about 90,000 through excess births over deaths, a total growth rate of about 17 per cent as compared to 27 per cent in the decade of the fifties and 20 per cent in the decade of the forties. However, even a 17 per cent growth rate per decade would boost population to well over a million people by the turn of the next century. The solid line on Graph 1 illustrates this growth arithmetically, and the solid line on Graph 2 illustrates the growth trend on semi-log coordinates.

As is the case in the United States as a whole, population
is not evenly distributed throughout the five-county area. Almost
fifty-five per cent of the people of the urban region live within
the administrative boundaries of the six major nodes, and an additional 7 per cent in the smaller cities and towns of the region.
There is, however, considerable urban/rural difference county to
county as evidenced by Table 1. Even these figures do not properly
characterize population distribution in the area, however, in that
population is heavily clustered in and around the urban nodes.
Maps 12, 13, and 14 illustrate this clustering in several different
ways.

Map 12 symbolizes residency with dots. Each dot represents
the residence location of 200 people. Clustering within and around
the urban nodes is evident. The map shows the concentration of
people within administrative city boundaries and immediately outside
them.

Map 13 shows population density by means of isopleths. The
map was compiled in great detail on computer punch cards by the
North Carolina State Department of Administration and generalized to
indicate population density gradients. The high population density
core corresponding to each urban center is apparent on the map.
From each core population density diminishes from the urban node
toward the rural inter-city spaces. The fairly regular gradient
from urban to rural is evident. The densely populated urban nodes,
the fairly densely populated urban peripheral areas, and the rather
sparsely populated intervening rural areas characterize the landscape.

Map 14 shows generalized residential land use for the five-
county region. The map was compiled from individual city and county
planning reports which vary in detail and accuracy, but does offer

GRAPHS 1 and 2--Population Growth

Graph 1 indicates population growth for the study area, for the years 1940 to 1970, on arithmetic conditions. The solid line indicates growth for the total area and the dashed line for the urban places with the area. Graph 2 indicates population growth for the study area, for the years 1940 to 1970, on semi-logarithmic coordinates. The solid line indicates growth for the total area and the dashed line for the urban places within the area. Both graphs show the very high total growth rates for the decade 1950 to 1960 and the even higher urban growth rates for the same period. Both graphs show the still high, but declining growth rates for the total area for the decade 1960 to 1970 and the declining urban growth rates for the same period. During the decade 1960 to 1970 the graphs show that total growth rate was higher than the urban growth rate. This must indicate a "filling-in" of the rural spaces between the urban nodes, although other evidence points to the fact that the process has not proceeded very far as yet.

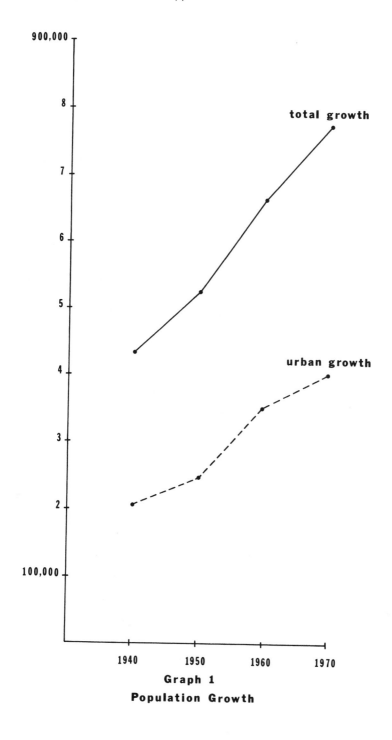

Graph 1
Population Growth

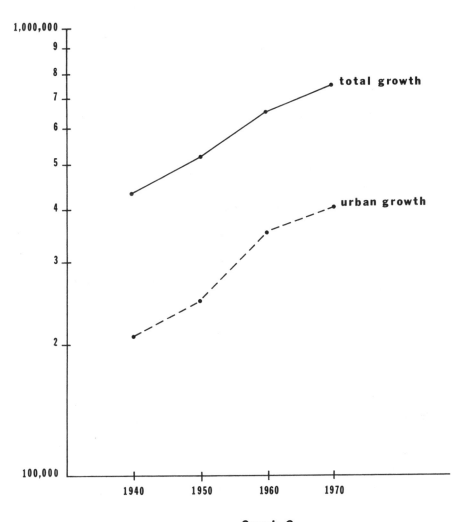

Graph 2
Population Growth

TABLE 1

COUNTY URBAN POPULATION 1970

County	Urban	% Urban
Alamance	50,497	52.4
Davidson	33,450	37.1
Forsyth	147,399	68.8
Guilford	220,127	76.3
Randolph	23,060	30.2
Total	476,533	61.78

MAP 12 - PIEDMONT URBAN AGGLOMERATION:
POPULATION DISTRIBUTION

The map symbolizes residency within the study area with dots.
Each dot on the map represents 200 people. It is evident that the
majority of the people living within the urban region live within
or close to the major urban nodes. The clustering of dots within
the administrative boundaries of Greensboro, Winston-Salem, and
High Point is evident, as is the clustering of dots, but not quite
so densely, within the administrative boundaries of Burlington/
Graham, Asheboro, and Lexington. The dots on the map are spaced
farther apart adjacent to but outside of the major urban nodes
and farthest apart in county peripheries. The map indicates the
clustering of people within and close to the major urban nodes
and the lower population densities farther away from the nodes,
with lower densities in county periphery.

81

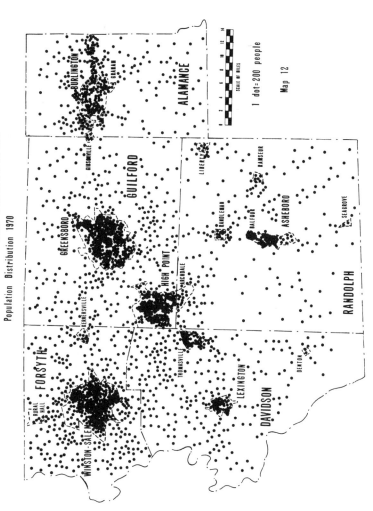

Piedmont Urban Agglomeration

Population Distribution 1970

SCALE OF MILES

1 dot = 200 people

Map 12

MAP 13 - PIEDMONT URBAN AGGLOMERATION:
POPULATION DENSITY

The map symbolizes three categories of population density with
isopleths: light, medium, and heavy. The high population density
core corresponding to each urban center is apparent. From each
core population density diminishes from the urban node toward the
inter-city spaces. The fairly regular density gradient from the
urban to rural is evident. The map indicates the discrete nature
of the urban nodes with regard to population density, but suggests
the beginnings of a "filling-in" process in inter-city rural space.
The filling-in process would seem to be taking the form of concentric zones outward from the nodes with sectors along high-speed
highways connecting the concentric zone areas. The map indicates,
however, that the filling-in process has not yet proceeded very
far.

Piedmont Urban Agglomeration
Population Density 1970

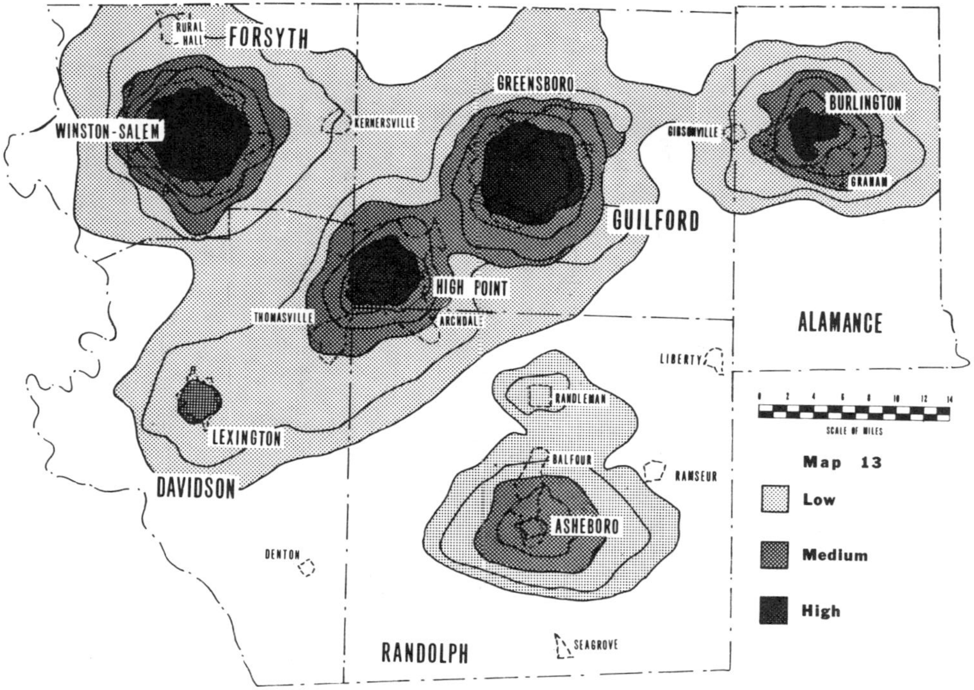
RURAL HALL
FORSYTH
WINSTON-SALEM
KERNERSVILLE
GREENSBORO
GIBSONVILLE
BURLINGTON
GRAHAM
GUILFORD
HIGH POINT
THOMASVILLE
ARCHDALE
ALAMANCE
LIBERTY
RANDLEMAN
LEXINGTON
DAVIDSON
BALFOUR
RAMSEUR
ASHEBORO
DENTON
RANDOLPH
SEAGROVE
0 2 4 6 8 10 12 14
SCALE OF MILES
Map 13
Low
Medium
High

84

MAP 14 - PIEDMONT URBAN AGGLOMERATION:
RESIDENTIAL LAND USE

The map shows generalized residential land use for the study
area and illustrates the clustering of people within and close to
the urban nodes and the light residential occupance farther from
the nodes. The map shows the concentration of residential land
use within city administrative boundaries and close to the major
urban nodes. Residential land uses decline from the nodes until
at the county boundaries virtually none of the land is residential.
As is the case with Map 13, Map 14 suggests the filling-in process
between the nodes, but shows that the process has not proceeded
far as yet.

Piedmont Urban Agglomeration

Residential Land Use

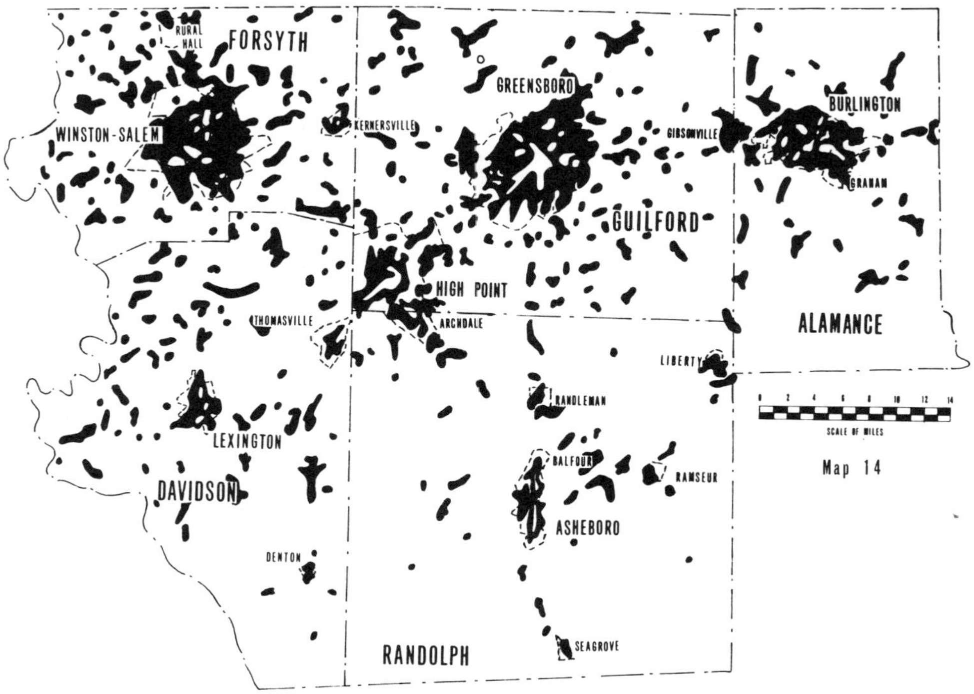
RURAL HALL
FORSYTH
WINSTON-SALEM
KERNERSVILLE
GREENSBORO
GIBSONVILLE
BURLINGTON
GRAHAM
GUILFORD
HIGH POINT
ARCHDALE
THOMASVILLE
ALAMANCE
LIBERTY
RANDLEMAN
LEXINGTON
BALFOUR
RAMSEUR
DAVIDSON
ASHEBORO
DENTON
RANDOLPH
SEAGROVE
0 2 4 6 8 10 12 14
SCALE OF MILES

Map 14

a general picture of urbanization within the region. The map shows
the concentration of residential land use within city boundaries
and in urban peripheries. The areas of residential land use become
smaller and farther apart from the urban nodes until almost no residential land use is visible in the far rural reaches of each county.

Map 12, "Population Distribution," Map 13, "Population Density,"
and Map 14, "Residential Land Use" all tell the same story. There
are more people living at higher densities using more residential
land at nodal centers than elsewhere. Each urban node contains a
high density core surrounded by zones of lower densities with lowest
densities in the inter-city spaces. Even though the several urban
nodes in the area may very well function as a single urban unit,
they appear to be discrete cities according to the criteria of population density gradients. Graph 3 reinforces this proposition.
The population density gradients of Greensboro, Charlotte, Fayetteville, and Raleigh with 1970 populations of 144,076; 241,178;
53,510; and 121,577 respectively have been plotted on the graph.
Of the four, only Greensboro could conceivably be a unit in a multi-
nodal urban system. Yet, the four density gradients do not differ
significantly. It appears that the nodes of the Piedmont region
do not differ significantly from other comparable size cities with
regard to population distribution and, as has been noted earlier,
with regard to population growth over the past three decades.

It has been pointed out that the decade of the fifties was
the period of peak over-all population growth for the study area.
This also was the decade of major nodal urban growth in the area,
although the urban population of the area grew more rapidly than its
total population for both decades 1940 to 1960. During the fifties
nearly three-fourths of the total increase in population occurred
in the six major urban nodes. It also has been shown that total

population growth in the sixties declined from the decades of the forties and fifties, and population growth also declined in the six major urban nodes in the sixties. The major nodes actually grew more slowly than did the area as a whole during that period. This would seem to indicate a filling-in of the open spaces around and between the major urban nodes during the decade of the sixties. Still, Maps 12, 13, and 14 show that the filling-in process is not yet substantially developed. Once the interstitial non-urban spaces of the Piedmont urban complex fill, of course, much of the present structure of the area will be changed, and its character as a "dispersed city" will have been transformed and perhaps lost forever.

Population projections for the next three decades have been developed by the United States Office of Business Economics for the counties comprising the study area. According to these projections, the growth rates for all counties except Randolph will be somewhat greater than the rate of the sixties, culminating in a total population of 1.2 million by the turn of the next century. A population of 1.2 million will surely result in extreme traffic congestion if an alternate to the automobile-oriented transport system of the area is not developed over the next three decades.

Two groups within the area have become involved with population planning for the multi-nodal region. The two protagonists of the case are the Piedmont Triad Council of Governments (COG) and the Greensboro Chamber of Commerce. The views of these two organizations can be rather neatly dichotomized into "population control" vs. "growth is inevitable". COG and the Chamber are the advocates of these two opposed points-of-view.

GRAPH 3 - POPULATION DENSITY GRADIENTS

The graph shows the population density gradients of four
cities within the state of North Carolina: Greensboro, Charlotte,
Fayetteville, and Raleigh with 1970 populations of 144,076;
241,178; 53,510; and 121,577 respectively. Of the four, only
Greensboro could conceivably be a unit in a multi-nodal urban
system. To facilitate comparison of the four density gradients,
they are not labeled with the city name, and all (some arbitrarily)
are terminated at the six-mile mark. It is obvious that the population gradients are all nearly the same, though not identical.
All gradients are rather high within a half mile of the city center, but decrease rapidly to a distance of 2 miles from city center. The density gradients for all four cities are almost identical
between 2 1/2 and 3 1/2 miles from city center, but differ somewhat
between 1 and 2 1/2 miles and beyond 4 miles. Nevertheless, the
graph indicates that the population density gradients do not differ
radically from 0 to 6 miles from city center; and it is, therefore,
suggested that units of a multi-nodal urban system do not differ
in this regard from comparable sized discrete single-centered cities.

89

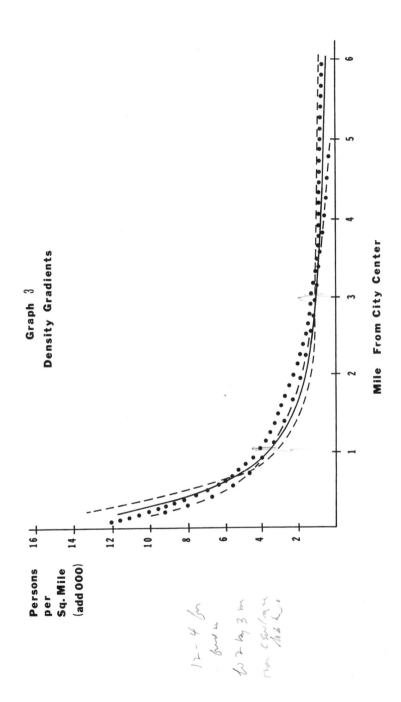

Graph 3
Density Gradients

The COG Position

The Piedmont Triad Council of Governments was formed in 1969
at the insistence of the United States Department of Housing and
Urban Development (HUD), to coordinate local planning efforts in
an eleven-county area which incorporates the five counties of immediate concern to this study. The Guilford County Planning Director
was appointed Executive Director of COG, and reports began to appear
in 1971. Local governmental units may choose to join or choose
not to join the Council of Governments, although all the major
governmental units have joined the council as have several minor
governmental units. The COG, acting under a directive, may solicit
local governmental membership from the five counties comprising
the study area plus six predominantly rural counties to the north
and east of the area. COG depends on local acceptance for enforce-
ment of recommendations or suggestions. However, COG retains a
rather large staff of professional planners and has employed consultants freely over the past few years. COG recommendations
for the development of the area are, therefore, available.

The COG stand on population growth for the five-county area
is simple. COG plans for a constantly declining population growth
rate with a goal of equilibrium at 1.1 million persons by the year
2000. It is COG's position that a population of this size will not
overload facilities as they are improved, in a realistic fashion,
over the next three decades. COG points out that present sources
of water are already adequate for the projected population for the
year 2000. COG is currently studying sewer and solid waste management and is confident that these systems will be ready for use as
they are needed. The transportation plan is rather vague, but suggests interstate standards for present highways linking the major
nodes of the urban system, supplemented by a segment of a state

(Charlotte-to-Raleigh) rail rapid transit line along the existing rail right-of-way. The rapid transit system would link Lexington, High Point, Greensboro, and Burlington along the main line, with a spur from Greensboro to Winston-Salem. The plan suggests improvement of existing airports and additional "strategically located" general aviation airports near any city of 10,000 people. Heliports and intracity buses would round out the transportation system.

The Council of Governments proposes to control the population growth of the five-county area by controlling industrial in-migration. COG points out that the geographic literature is rich in papers linking population growth to the industrial growth in the southern Piedmont. Specific controls to industrial in-migration have not been formulated, but they night be a combination of restrictive zoning, restricted utility services, stiff pollution control regulations for new industry, and tax policy. The Council of Governments sees no real problem in restricting industrial in-migration, if the support of the people and the support of the governmental units of the region can be obtained.

The Council believes that control of the distribution of population is even more important than control of growth and numbers, and it suggests that the multi-nodal character of the area be retained. COG argues that Greensboro and Winston-Salem should not be permitted to grow beyond a certain population level, perhaps 250,000; that Burlington, High Point, and Asheboro should be held to smaller populations, perhaps 100,000; and that Lexington and existing smaller cities should contain even fewer people, perhaps only 50,000 each. Population over-flow will be diverted to planned "new towns" located within easy auto commuting distances from Greensboro, Winston-Salem, and High Point and to existing smaller cities and towns within the region. In addition, large "green belt" management areas between urban nodes will preclude obliteration of the physically discrete character of the urban centers. COG believes that urban size, density, and form can be controlled through area-wide zoning, through city boundary control, and through the controlled distribution of utilities and city services.

Planned management green belts present a more difficult problem, according to COG. Some of this land can be purchased and converted to public parks. The bulk of the land, however, must be kept "open" by other means. Area-wide zoning, utility extension policy, tax credits for open land, and the purchase of easements are the methods suggested by COG for retention of green belts.

The Chamber Position

The Greensboro Chamber of Commerce, though organized and financed in the typical chamber fashion, is a cut above the usual organization devoted to the needs of local businessmen. The Greensboro Chamber maintains an active urban research division that has produced several sophisticated publications concerning local and area-wide problems. As is the case with COG, both staff and consultants have contributed to the chamber efforts. The Greensboro Chamber of Commerce perceives Greensboro as a unit in a larger city system and is concerned with area-wide urban problems as well as with local urban problems.

The Greensboro Chamber has no objection to the concept of population growth control, but it objects to the suggested COG methodology for that growth control. The COG plan to limit population growth by restricting industrial in-migration is anathema to the Chamber. The Chamber points out that only about 20,000 people were added to the urban complex during the past decade by in-migration, while almost 90,000 people were added

through an excess of births over deaths. The Chamber believes
that to limit jobs through the control of industrial in-migration
would force the out-migration of the working age natives, in the
Chamber's words: "the contributors – the cream of the crop."
The Chamber is adamantly opposed to any policy that might encourage this possibility.

The Chamber claims it does not promote growth but believes
growth preferable to any suggested alternative and thus inevitable.
The Chamber suggests that a 19 per cent average population growth
rate per decade is reasonable and realistic. At this rate the
Piedmont urban complex would reach the 1.1 million mark by 1990
and would total 1.35 million inhabitants by the year 2000. The
Chamber agrees with COG that planning, including transportation
planning, for 1.1 million people for the five-county area poses
no great problem and suggests that a population of this size can
be accommodated by 1990. The Chamber position concerning population growth beyond the 1.1 million mark to the year 2000 and
beyond is that by such a time technological breakthroughs will be
available to resolve problems that would seem to defy a solution
today. Who can say, according to the Chamber, what technology
for water distribution, sanitation, pollution control, and trans-
portion will be available by 1990? At any rate, according to
the Chamber, the year 1990 is soon enough to apply population
control should the need arise by then.

The Greensboro Chamber of Commerce agrees with COG that population distribution is more important than numbers. The Chamber
believes, however, that migration will, inevitably, be into the
three largest cities: Greensboro, Winston-Salem, and High Point.
The Chamber believes this is desirable because the proper infrastructure

for accommodating large populations is already available in these
cities and can readily be expanded to meet future needs and to control distribution within these urban units. The Chamber believes
that the physically discrete nature of the urban units of the urban
complex can be maintained by the development of new towns and by
limiting the population of the existing smaller cities and towns
in the area. Growth and distribution control is, according to the
Chamber, proper for the smaller urban units because the infrastructure is not available in small cities and cannot conceivably
be developed in time to control sprawl. It appears that the Chamber
considers "sprawl" inevitable, perhaps even desirable, for the larger urban units but undesirable for the smaller cities and towns of
the area.

The Chamber would be delighted with a ring of planned management green belts, but considers the concept a "pipe dream" beyond
the small amount of land actually purchased. The Chamber points out
that zoning has consistently failed in the courts if there is a
higher and better use for a large block of land. The Chamber has
considered the possibility of offering tax credits to farmers in
the inter-city spaces who do not convert land to urban uses and has
considered the plan "doomed to failure". According to the Chamber,
each farmer would accept the credits until offered a high price to
convert the land to urban uses. The Chamber predicts that the
farmer would then sell the land to the highest bidder. The Chamber
has considered the possibility of purchasing 99-year "easements"
that require individual farmers to retain land in rural uses, and
also pronounces the plan "doomed to failure". According to the
Chamber, a farmer in an urbanizing county would be a "fool" to
accept a small annual easement payment with potential riches
"around the corner". Large payments are not economically feasible,

and the Chamber does not consider the area farmers "fools." The
Chamber believes that utility extension policy does not control
development; it only controls the direction or location of development. This too, according to the Chamber, will fail as a method
to preserve inter-city "greenbelts."

The two opposed points-of-view with regard to population planning, as expressed by the Piedmont Triad Council of Governments
and the Greensboro Chamber of Commerce, probably represent a dicho-
tomy between the administrative bureaucracy and the typical resident of the area. This statement cannot be readily quantified,
but interviews with mayors, councilmen, and planners of the area
indicated a positive attitude toward continued population growth as
symbolized by the Chamber position. On the other hand, although no
resident whom we interviewed mentioned COG as a governmental unit
affecting his life and although less than ten per cent of the interview sample could identify COG even vaguely, respondents indicate
by their answers to a multitude of questions that they favor the COG
position. This can be summarized by the factual statement that most
residents of the area like living in a "dispersed city". They like
it because of its dispersed-city character, and they want to keep
it that way. This statement is examined in greater detail in Chapter V. COG might promote the "population control" concept with more
success if the message could be delivered to the people instead of
to the bureaucracy. The first task of the Council of Governments,
however, is to let the residents of the Piedmont Dispersed City know
that a governmental unit such as COG exists.

The COG Publications

Regional planning, if the COG publications are indicative, is
in the very early or developmental stage. Published COG studies
began to appear in 1971 and some 50 publications are available from

COG headquarters. (The COG published bibliography is included in the appendix.) The general tone of the studies is one of despair over lack of data. This obstacle to planning is evident in the following quote from the Solid Waste Study:[1]

> "Without a substantial effort to establish a uniform base of information, few conclusions can be reached at this time. Even the dimensions of the issue itself are clouded by the shortage of available facts."

The COG offers publications covering such diverse subjects as transportation, codes, water supply and disposal, housing, open space/recreation, criminal justice, environmental affairs, manpower, health, libraries, and development, in addition to solid waste management. The reports all consist of statements concerning the current "state of the art" in the COG region and a listing of data (or other) needs. All decry the lack of information necessary to the formulation of plans. For example, the solid waste studies further suggest a program of systematic data collection so that a solid waste management plan can be prepared.[2] The plan, according to the report, when prepared should include the following:

   a. Identification of land requirements for disposal sites

   b. Recommendation of site locations

   c. Suggestion of population densities suitable for inclusion of service districts

   d. Institution of an information and monitoring system on regional solid waste management data

   e. Provision of support for State and Federal agencies

   f. A clearinghouse for technical information

   g. Recommended alternatives

[1]The Research Group, Inc., Regional Solid Waste Disposal Study-Phase I (Greensboro, N.C., Piedmont Triad Council of Governments, April 1972), pp. 14-15.

[2]Ibid., Phase I, p. 24 and Phase II, pp. 49-51.

Chapter V indicates that the typical resident of the urban complex rates most local governmental services at least "adequate".
The primary exception to this rating, as determined by the resident
interview survey described in Chapter V, is solid waste collection.
All residents who depend on private collectors rate the service
as "poor". All residents who must dispose of solid waste themselves
blame county government for not providing the service. It is evident
that the COG must formulate and implement a solid waste management
plan soon if it is ever to become a representative government for
the region.

Chapter V further indicates that inter- and intra-nodal travel
difficulty is the focus of citizen discontent within the urban region
although residents do not associate transportation with governmental
services. It is fortunate for the Council of Governments that the
"man on the street" does not blame his government for travel fric-
tions for COG has not yet scratched the surface of transport needs.
The COG publication that considers transportation merely states some
of the transportation problems in the COG region and suggests a
series of broad goals and objectives covering the total transportation system and the sub-systems of rail, highways, motor freight
carriers, motor passenger carriers, airports, terminals, and associated land use.[1] The entire list of goals and objectives is too
lengthy to be totally reproduced here, but the goals reproduced
below for the total transportation system offer an indication of
the very general tone of the publication:

"Near-term goal: Develop a coordinated and integrated
multi-modal transportation system consistent with the economic, population, land use, environmental, and regional
requirements of the state."

"Long-range goal: Provide for the development of new
techniques in transportation modes or sub-systems to satis-
fy special transportation problems, overcome existing and
future deficiencies, and assure the development of a balanced
transportation network."

[1]Piedmont Triad Council of Governments, The 1974 National Transportation Study Narrative for the Units of Government in Region G,
(Greensboro, N.C., August, 1973), pp. 1-12.

It is unfortunate that the COG has not been able to develop a transportation plan for the area, for transportation is surely the crux of any "dispersed-city" system. Without a reasonably efficient intra- and inter- nodal transport system beyond the private automobile, increasing population pressure will soon make it very difficult to live in one node, shop in another, work in a third, seek professional services in yet another, and so on.

Reports on water resources, housing, justice, open space, environmental affairs, health, man power, and libraries all follow the theme of solid-waste and transportation studies. The reports all consist of the "current state of the art", a listing of the problems, a suggestion for data collection, and a very broad statement of goals and objectives. The statements reviewed in this chapter (The COG position) concerning planning and development are the closest effort to a "plan" that COG has to offer. Even these statements are quite general and must be supplemented with interviews with the COG staff to be meaningful. Perhaps it is too soon to expect "plans" from a governmental unit only a few years old, or perhaps the fact that COG can only act with the permission of its member governments precludes the unanimity of approval necessary for plan formulation and implementation.

Other Planning Suggestions

Greensboro completed its "Gateways" community project in February, 1974. The project was designed to involve the people of Greensboro and Guilford County, including elected and appointed government officials, in two all-day discussion sessions concerning local problems. The first session conducted in December, 1973 was preceded by the perusal by all interested persons of "position papers" on several subjects, including "community development". The community development position paper stressed the Greensboro "relative location".[1] The paper discussed Greensboro as one urban node in a larger city system referred to as "The Piedmont Dispersed City". It pointed out the inter-nodal interaction of the urban region and suggested that nodal specialization might be the reason for this interaction. The paper proposed that this urban form might be a pleasant way to live happily in a city environment, but cautioned that inter-nodal travel is the crux of dispersed-city organization and pointed out that traffic friction is thought by the resident to be an ever-increasing problem. The paper concluded that the "dispersed-city" character or "relative location" might be the primary resource available to the people of Greensboro.

It is impossible to state whether any benefit will be derived from the "Gateways Project." However, one point was made with a number of people, including a few city and county officials, that Greensboro is, indeed, a unit in a larger urban system and depends, in part, on the other units of the system. Several "Gateways" participants suggested at the December, 1973 or the February, 1974 meetings that Greensboro residents must begin to think in terms of a "dispersed city".

Summary

This chapter has pointed out that over-all population growth in the study area reflects national trends for the three decades, 1940 to 1970. Further, the chapter suggests that each urban node of the city system does not differ significantly from other comparable sized North Carolina cities with regard to population density gradients. Over-all population growth and distribution, then, do not help distinguish the "dispersed-city" character of the area. However, population growth and distribution do point up the

---

[1]Charles R. Hayes, Community Development (Greensboro, N.C. Gateways, 1973), pp. 2-8.

need for over-all planning for the urban region. Resident interviews (Chapter V) indicate dissatisfaction with sanitation services, with transportation services, and with shopping facilities and indicate satisfaction with the "dispersed-city" character of the urban region. It would seem that the population of the area has reached the level where planning for these services is necessary before further population growth takes place. If further population growth begins to obliterate the discrete character of the urban nodes, the current satisfaction with the "dispersed-city" character will surely be replaced by general dissatisfaction and perhaps by specific dissatisfaction with the governmental units of the urban region.

The agency most likely to succeed with an over-all planning effort is the Piedmont Triad Council of Governments. The planning efforts of this governmental unit have, to date, been ineffective. The average citizen does not know of the Council's existence; the average bureaucrat resists the Council's planning suggestions; and the Council's published plans are too vague and general for implementation. Nevertheless, if the COG can find a way to reach individual residents with its population growth and distribution plan, it would discover that the typical resident agrees in general with the suggestions in these plans. If the citizenry approves the plans, bureaucracy cannot lag far behind. Lack of data for specific plan implementation is a larger problem, but perhaps it too could be solved.

Chapter V

ON THE PERCEPTION OF A DISPERSED CITY

"It has been said that beauty is in the eye of the
beholder. As a hypothesis about localization of function,
the statement is not quite right - the brain and not the
eye is surely the most important organ involved. Nevertheless, it points clearly enough toward the central problem of cognition. Whether beautiful or ugly or just conven-
iently at hand, the world of experience is produced by the
man who experiences it."[1]

Perception and Cognition

There certainly is a real world of buildings, streets, trees,

grass, and even books, but we have no direct access to these things.

Whatever we know about the world around us has been perceived by the

senses and interpreted by the brain. The stimuli that may result

in an item of experience bear little resemblance either to the real

object or to the experience the perceiver will construct.[2] The

chain of events producing the experience has been termed "perception-

cognition-conceptual behavior." Perception is the reception of the

stimulus, cognition is knowing or believing, and conceptual behavior

is the ultimate use to which the information is put.

Perception studies have been multi-disciplinary. Psychologists,

sociologists, anthropologists, planners, architects, philosophers,

---

[1]Ulric Neisser, Cognitive Psychology (New York, Meredith
Publishing Company, 1967), p. 3.

[2]Ibid., p. 3.

and geographers all have been interested in perception-cognition-conceptual behavior. The geographer, however, has been interested primarily in the phenomenological approach to perception. This approach views the environment not as it is, necessarily, but as it seems to be to the cognizing individual. As man interacts with his environment, he behaves in accordance with what he believes the environment to be, not necessarily what it is. Since error can creep into any stage of the cognizing process, the perceived environment may not be a replication of the real world.

It might seem that the phenomenology of environmental perception would produce as many different views as there are perceivers, since no two people can occupy the same place in the environment at the same time. Even if time and space could be identical for two participants, the environment of each would be unique since each participant would reflect his own interpretation of the immediate environment and his interrelation with other participants. Yet, there is evidence that not only can there be group perception but behavioral continuity; that is, not only can members of a group perceive the environment similarly, but there may be patterns of behavior in response to the environment that persist regardless of the individuals involved.[1] Group perception and behavioral continuity are, of course, strongly reinforced by learned behavior patterns.

Perception of even a portion of the environment is a wholistic process, but the whole of a complex object is too much for most people. The perceiver immediately starts an abstracting process. Major features of the perceived object are abstracted as an aid to the next step in the process. The fact that the object remains visible - in the mind's eye, so to speak - aids in the pro-

---

[1] H. M. Prohansky, W.H. Ittleson, & L. G. Rivlin, "The Influence of the Physical Environment on Behavior: Some Basic Assumptions" in Idem, Environmental Psychology (New York; Holt, Rinehart, and Winston, 1970), pp. 27-37.

cess of abstraction. This ability to see an object for as long as one-tenth of a second after removal of the stimulus has been termed, "iconic memory."[1]

In order for the perceived information to be used by the individual - now or later - it must be committed to "short-term store". This is the portion of human memory that stores information temporarily for either immediate use or transfer to the permanent memory (long-term store). Storage of information in short-term store is accomplished by grouping and rehearsal. Experimentation has demonstrated that if subjects are not permitted to rehearse, it is virtually impossible for them to commit data to short-term store. Since short-term store has a very limited capacity, grouping is a device used to extend that capacity. For example, if a person can normally commit seven digits to short-term store, groups of threes will extend the capacity to twenty-one digits. Information committed to short-term store is retrieved for use or further processed for long-term store.

Information destined for long-term store, or what generally would be called memory, is put through a coding process. Data to be committed to long-term store are coded visually or linguistically. Generally, a code is selected by the individual to facilitate retrieval; that is, the individual commits the information to long-term store the way he thinks he will retrieve and use the information later. Whereas short-term store is limited in both duration and capacity, long-term store is seemingly unlimited in both dimensions. However, when error creeps into the cognitive process, it is more often due to memory failure than to perceptual error.

Finally, information to be used must be retrieved from long-term store and returned to short-term store. At this point the

[1]The paragraph above and the following three paragraphs have been synthesized from Neisser, op. cit. pp. 1-10.

information is ready for use for the final step of the process: conceptual behavior.

Many environmental perception studies have been and are being conducted in several disciplines, including geography.[1] These studies tend to concentrate on examples of group perception and behavioral continuity rather than on developing laws to explain these phenomena. This study is no exception.

Function of the Chapter

The study area is described in this chapter in terms of resident perceptions. This had been accomplished through analysis of answers to the following questions:

1) Do residents like living in a dispersed city environment, why or why not?

2) Do residents perceive inter-nodal specialization; if so, in what functions?

3) Do residents want a common government for the area; if not, why not?

4) Is the "idea of city" developed among residents of the area?

Of course, the interview questions were phrased in a much different manner with as many as eight or ten specific questions asked of respondents in order to elicit answers to the four major questions. A fifth major question, prompted by the Burton article,[2] was asked though it has nothing to do with perception:

5) Do residents leave the area for "top-level" shopping an unusual number of times during the year?

The analysis of the resident answers to this question is included in this chapter, rather than elsewhere, because the answers were

---

[1] See, for example, Thomas F. Saarinen, "Perception of Environment," Association of American Geographers, Commission on College Geography, Resource Paper No. 5 (1969) -and- H. C. Brookfield, "Environmental Perception" in Progress in Geography, Vol. I ed. Board, Chorley, Haggert, and Stoddart (London, Arnold Press, 1969), for some two hundred references on perception related to geography.

[2] Burton, "A Restatement, etc." op. cit., p. 288.

obtained through interviews.

This chapter, then, is primarily devoted to the consideration of the perception, cognition, and resulting behavior of the residents of the Piedmont urban complex, or how the residents "feel" about their environment and what do they do in response to their feelings. Secondarily, the chapter is devoted to an actual analysis of resident trips beyond study area boundaries.

The Resident Sample

The resident sample was composed of a two-hundred person spatially random sub-sample of the total area and one-hundred fifty person spatially random sub-sample from the major urban nodes: thirty each from Greensboro and Winston-Salem; twenty-five each from High Point and Asheboro; and twenty each from Burlington and Lexington. The sample breakdown is as follows: forty-eight per cent urban residents (residents of the six major nodes) and fifty-two per cent rural or small town residents; fifty-five per cent females and forty-five per cent males; sixty-nine per cent white residents and thirty-one per cent non-white residents (all non-white residents were judged to be Negro); ten per cent affluent residents, seventy-five per cent middle-income residents, and fifteen per cent low income residents; and twenty-per cent over age 60, fifty per cent between the ages of 40 and 60, thirty per cent between the ages of 20 and 40, and no one under age 20. Race and sex were judged by the interviewer, affluence was based on a rough appraisal value of the residence by the interviewer, and age category was judged by the interviewer.

Spatial randomness was obtained through the use of map grids with 1,000-square-foot cells selected with a table of random numbers. Interviewers were instructed to approach the residence closest to the center of the appropriate grid cell, but if a successful inter-

view could not be obtained, then to approach the next closest and
so on until an interview could or could not be completed. If no
interview could be obtained in a particular cell, it was discarded
and another cell was selected at random.

Interviews for the spatially random sample of the entire
area were obtained during March, April, and May of 1971. Interviews for the spatially random sample of the six major urban nodes
were obtained in March, April, and May of 1973. Each interview
consumed about an hour and interviewers reported no unusual problems
in obtaining satisfactory responses. A copy of the questionnaire
is reproduced in the appendix. The reader is, however, urged to
approach the questionnaire with caution since interviewers were
instructed to obtain as much depth as possible, that is, respondents were encouraged to "speak their minds," while interviewers
made notes on the conversation for later transcription.

Perception of the Environment

Ninety-four per cent of the respondents would rather continue
to live in the study area than anywhere else in the world. Two
per cent would rather live in Florida, and one per cent each opted
for New England, the deep South, Appalachia, and the West Coast.
The tongue-in-cheek prediction one hears, that by the year 2,000
the entire population of the United States will reside on the west
coast, has failed to consider Piedmont North Carolina. It is true
that sixty per cent of the respondents were born in the area, but
most of the in-migrants are apparently converted Carolinians. The
residents who would move west and southwest were returning to the
scene of their childhood; those interested in Florida and New England
were seeking climates more to their liking.

Just because nearly all the respondents preferred life in
the area to life anywhere else does not confer the accolade

of "optimum environment" on the area. However, the reasons offered
in defense of the good life tend to support that possibility
Residents liked the very things that tend to make the place a "dispersed city." Respondents mentioned the combination urban/rural
atmosphere -- the potential for being a few minutes away from the
country in the city and vice-versa. They mentioned the potential
for shopping in several central business districts or in several
shopping centers in different urban nodes. They considered the
possibility of rural living and city work (and/or vice/versa) a
plus factor. They like the potential for a 'quiet' life yet a life
with urban amenities available. Responses may not indicate opti-
mality, but they surely point toward the superlative.

It must also be pointed out that the three hundred fifty
respondents also repeatedly mentioned three advantages to living
in the area that have nothing to do with dispersed-city characteristics.
The friendliness of the local people was pointed out fifty-four
times. It seems a reasonable a priori assumption that friendly and
unfriendly people can be found in nearly every city of the world.
Pleasant climate was mentioned thirty-eight times. Although Terjung
refers to the North Carolina Piedmont as oppressive in summer with
a compensating mild winter and with pleasant transitional seasons,[1]
the "pleasantness" of a climate is probably closely associated with
the predilections of the person experiencing it. The beauty of the
scenery was reported thirty times. Although many people do consider
the southeastern Piedmont quite scenic, with its randomly rolling
hills and year-round greenery, beauty is surely in the eye of the
beholder.

---

[1] Werner H. Terjung, "Physiologic Climates of the Conterminous
United States: A Bioclimatic Classification Based on Man," Annals
of the Association of American Geographers, Vol. 56, No. 1 (March
1966), pp. 141-179.

Most of the respondents find the area to their liking and like the very things about it that make it a "dispersed city". Everyone, though, can find something to dislike about any place. The interviewers were instructed to discover these dislikes through questioning.

Travel difficulties led the list of dislikes. Twenty-five per cent of the respondents mentioned the difficulties of travel without prompting, and one hundred per cent of the respondents reported travel difficulty as a problem when led by the interviewer. Further, all respondents believed travel difficulties to be increasing. Climatic dislikes, specifically "too hot in summer and/or too cold in winter" were second with references by sixteen per cent of the sample. Lack of big city amenities was a close third, having been mentioned by 15 per cent of the respondents. A scattering of responses covering unfriendly people, the high cost of living, inadequate governmental services, and poor soil (from a farmer) rounded out the list of dislikes. Many residents, however, could find nothing except the travel friction to dislike about the area except a fear that in the future the area would lose its "dispersed city" character. Although the term "dispersed city" was not used by any interviewer or interviewee, the responses to several questions concerning the possibility of specific urban nodes "growing together" could only be interpreted as relating to the characteristics of a "dispersed city".

The typical resident of the area not only likes living in a "dispersed city" environment, but he also wants to keep it that way. More than half the respondents thought the urban nodes of the area would grow together to form some sort of minor megalopolis within 20 years; all but 18 per cent thought it would happen some day. All respondents deplored this possibility. Half the respondents, though deploring the impending spread of the urbanized areas, considered it inevitable. The other half, however, had suggestions to avoid the possibility. In fact, the 18 per cent who considered the possibility remote believed something official would be done to block urban sprawl.

Twenty-five per cent of the total number of subjects (that is, half of the half who believe urban consolidation avoidable) linked population growth to industrial growth.[1] These respondents believe, apparently accurately, that industrial, particularly manufacturing, growth is the primary causal factor in population growth in the area. They suggest that industrial growth be curtailed through governmental action, even though they recognize that such action may widen the wage and salary differential between this area and other more prosperous areas of the county. It must be concluded that a man willing to "tighten his own belt" must believe he has something to protect.

Perception of Retail Specialization

Retail specialization has been cited as a key characteristic of dispersed city functions. The dispersed city consumer has a choice of several shopping places for retail goods at all levels, but specialization in many other functions is also possible. Manufacturing, entertainment, professional services such as medicine, advertising, public relations, banking, higher education, and wholesaling are all possible nodal specialities. Although it is doubtful that inter-nodal specialization actually exists in the study area, for the purpose here, it matters only that the resident believes it to be so. Do the residents of this area, as a group, perceive nodal specialization as a function of the Piedmont urban complex? The survey indicates the affirmative.

Two-thirds of the respondents indicated by their responses to a group of questions that they did perceive nodal specialization

---

[1]See Chapter III for references pertaining to the linkage of population and industrial growth.

within the area. Those subjects believed it to occur in several ways.
Retail services, entertainment, manufacturing, and professional services were the primary kinds of specialization according to the survey
results. However, a third of the respondents did not believe specialization occurs.

The perception of specialization in retail services was
investigated through the survey. The sixty-six per cent who believed
retail specialization obtained, pretty well agreed on the image projected by the major retail zones in the area. This is an example
of group perception.

Winston-Salem and Greensboro are about the same population
size, and their downtown trade areas are very close to the same size.
The number of retail core functions, range of goods, range of prices
do not differ greatly, but the two downtown areas have vastly different images as perceived by the people who shop in both central
business districts.

The Greensboro central business district is perceived as the
"action" center and the Winston-Salem Downtown as the "passive" center; that is, Greensboro is perceived to have more shoppers at any
given time and a greater range of goods but higher prices. Greensboro is also perceived as the style center for the Dispersed City,
particularly for women's clothing.

The Asheboro central business district is smaller than the
Winston-Salem or Greensboro districts in number of square feet of
retail sales area and number of retail functions. The Asheboro
downtown trade area, however, is comparable in size to the two
others. For example, the ninety-nine per cent confidence limit for
the Asheboro trade area is 23.87 miles as compared to Winston-Salem
at 27.71 miles and Greensboro at 23.75 miles. The Asheboro downtown
area obviously offers something of value to the Dispersed City
resident. Asheboro's downtown is perceived as the comfortable or

"folksy" central business district. Merchants are perceived as friendly and helpful. Parking is thought to be convenient and inexpensive, and prices of goods and services are perceived as cheaper than in Winston-Salem and in Greensboro. Dress is seen to be informal. The friendly image has extended Asheboro's downtown trade area beyond its expected reach considering the relatively small population of the city and the relatively limited number of retail functions within the central business district.

The High Point central business district projects a negative image to the Dispersed City resident. Prices, parking difficulty, and traffic congestion are perceived to be comparable to Winston-Salem and Greensboro, but number of functions and range of goods are thought to be inferior to the two larger nodes. Trade area reach reflects this negative image. The High Point downtown 99 per cent confidence limit of the retail trade area is only 4.6 miles.

The Burlington and Lexington downtown areas seem to project a neutral image. Note on Map 4 (Chapter II) that neither downtown looms prominently as second shopping choice. Access and distance decay are the primary limiting factors for the two downtown cores because of the absence of magnetism linked to image. The 99 per cent confidence limits for the two trade areas are as follows: Burlington 17.7 miles and Lexington, 16.6 miles.

Shopping centers within the area also project certain specific images perceived as such by the resident. For example, regional center A in Greensboro, regional center B in Burlington, and regional center C in High Point are in the same size range as measured by square feet of buildings. Retail functions do not differ greatly, nor do range of goods and prices. Perception and use exhibit great differentiation.

Shopping Center A is preferred two to one over the other two centers according to survey responses. Examination of tabulation sheets would lead any investigator to the conclusion that center A commands the lion's share of the business, but in fact that is not the case. The trade areas for the three centers are approximately the same size. Further respondent questioning reveals that Center A is perceived as the upper middle and high-class shopping center. Therefore, retail offerings at Center A must be of a higher quality. The respondents might reason that they should all shop there, so they tell the interviewer that they do. Centers B and C, however, get as much of the business. Center B is perceived as the middle-class center, and not only do people feel more comfortable there, but they really believe prices are lower. Center C is perceived as the middle and lower-middle class center and, as such, commands much of the business from middle and lower-middle income consumers.

An interesting sidelight concerning the perception of shopping center image was revealed through the questionnaire survey almost by accident. As the survey progressed, it became evident that the people living in a particular area of lower cost housing did not patronize the closest shopping center. The residential district had good access to the center and was potentially within the shopping-center trade area. Interviewers were sent into the low-cost housing area to try to determine why these people did not use the closest shopping center. In-depth questioning revealed that high-value housing surrounding the shopping center constituted a perceptual barrier that the lower income people were reluctant to cross.

The typical area resident believes that the various nodes within the urban agglomeration specialize in certain high-level retail offerings. Although no research results are available on this subject, observation would lead most people to the conclusion that the resident is sensitive to the image of "feel" of a specific node or the retail clusters within that node and transfers this feeling of difference or uniqueness to the goods or services purveyed by the commercial agglomeration. Perhaps each node does project an image all of its own. If so, specialization is a reality, though in a different sense than that of brand or style.

Perception of Media Advertising Specialization

Perception of specialization is also demonstrated through media advertising. A study of the influence of television, radio, and newspaper advertising in North Carolina revealed that television advertising has very little influence on shopping place.[1] Television viewers could recall national advertising but not place or store advertising. This does not imply that television advertising does not influence purchasing, only that it does not influence the place of purchase. This proposition is further substantiated by the fact that the coefficient of correlation between television service area reach and downtown trade area reach was quite low, .35. Newspaper advertising, on the other hand, is perceived as a guide to place or a store for retail purchase, and perhaps this perception is quite accurate. This does not mean that newspaper advertising is more effective than television advertising. The newspaper is simply a more effective medium than television for designating specific stores or shopping centers to the consumer. This consumer perception was substantiated by the fact that the coefficient correlation

[1]Bennett and Hayes, op. cit. pp. 44-52.

between newspaper service area reach and downtown trade area reach
was fairly high, .83.

Perception of Manufacturing Specialization

The resident's perception of specialization in manufacturing
in the area also is marked, at least for the three major urban nodes.
The respondents who believed nodal specialization to be a characteristic of the urban complex readily identified manufacturing
dominance in Greensboro, Winston-Salem, and High Point. Greensboro
is perceived, quite accurately, to specialize in textile and clothing
manufacture, although the rather large Greensboro manufacturing base devoted to machinery and electrical machinery is almost
ignored. Winston-Salem is perceived, again accurately, to specialize
in cigarette manufacture, although the electrical machinery compo-
nent of the Winston-Salem manufacturing base was scarcely mentioned.
High Point's national reputation in the manufacturing of furniture
is correctly assessed, but the large knitting mill complex was only
a small part of the respondent's cognition of the High Point manufacturing scene. The Greensboro manufacturing complex is based
primarily on textiles, clothing, and machinery; cigarettes and
electrical machinery are the most important components of the
Winston-Salem manufacturing base; and High Point does indeed produce furniture and knit-wear in abundance. Yet, to assume that these
nodes "specialize" in these products exceeds reality. All three
nodes contain textile mills, clothing factories, furniture manufacturing plants, and machinery manufacturing plants and Greensboro produces cigarettes. It must be concluded that inter-nodal
manufacturing specialization is perceived rather than actual.

Perception of Recreational Specialization

Greensboro is the "fun city" of the central North Carolina
Piedmont, if the perception of the residents is accurate. This

would seem to be due largely to a 15,500 seat coliseum and a few
night clubs. The service area of the Coliseum has never been accurately measured, but the function is well known and apparently well
attended by area residents. Whether or not the Coliseum service
extends beyond area boundaries is not known. The service areas of
the night clubs are not known either, but their clientele is mostly
transient.[1]

Restaurant dining is the favorite commercial recreational activity for Greensboro residents.[2] Since this is overwhelmingly true
for one node, it is likely true for all nodes. Winston-Salem is
the "restaurant capital" of the Piedmont urban complex. Winston-
Salem was mentioned sixty-five times as the node with the best
restaurants; Greensboro forty times, High Point eighteen times,
Burlington ten times, Lexington seven times, and Asheboro six times.

Perception of Professional Service Specialization

Specialization in certain professional services also is discerned by the residents of the area, and they believe that specialization occurs in advertising, insurance, and photography. Advertising
service is perceived as a function of the two largest nodes. Winston-
Salem and Greensboro are seen as equally competent in furnishing
advertising service. This is very likely a function of population
size; that is, without any real knowledge of the availability of
advertising services, the resident assumes the two largest nodes in
the area could be expected to furnish the service. Greensboro is
accurately perceived as the insurance center of the area. The node
does contain several important insurance companies, and the residents
are aware of this fact. High Point completes the triad with Winston-

---

[1]Hayes, Greensboro Recreation Survey, op. cit. pp. 1-10.

[2]Ibid., pp. 1-10.

Salem and Greensboro in the perception of the availability of industrial or commercial photographic services. Winston-Salem and Greensboro are, again, very probably perceived as likely places to offer
professional photographic services because of their population size.
High Point contains an important commercial furniture photographer,
and apparently the area residents are aware of this fact.

Perception of Specialization in Higher Education

Specialization in education is also perceived, but it is difficult to assess. The area contains three major universities and
several colleges and/or professional and technical schools. At
first, the interviewers attempted to assess resident perception of
specialization for all institutions of higher learning within the
area. Since they were not successful, the interviewers concentrated
on a state university, a private university, and a private liberal
arts college.

The University of North Carolina at Greensboro, Wake Forest
University in Winston-Salem, and Guilford College in Greensboro
were the institutions used as examples of higher education in the
region. Respondents were questioned in depth regarding their perception of the images projected by these three institutions.

The University of North Carolina at Greensboro presents a
vague image to most residents. It was often referred to as "Women's
College" although it has been coeducational for more than a decade.
Respondents know of its existence and correctly located the institution in Greensboro but could offer little else to characterize
the institution. Respondents believe it was once a very fine women's
college, but could offer few current distinguishing characteristics.
Enrollment was consistently underestimated, although instructional
quality was rated "very good". Respondents, however, had very little
knowledge of college, school, or course offerings within the univer-

sity.

Wake Forest University in Winston-Salem projects a more definite image than does UNC/G. This seems to be due largely to the presence of a medical and a law school and a "big time" sports program. Enrollment was consistently overestimated and instructional quality was rated "excellent". Respondents know of the medical school and were very familiar with the Wake Forest football and basketball teams.

Guilford College in Greensboro was also known to the respondents. The perception of the institution was one of a formerly very high ranking institution academically that had declined in academic excellence over the past several decades. Instructional quality was rated below UNC/G and Wake Forest but still "good". Respondents were aware of the church control of the institution and that offerings were primarily in liberal arts and religion.

In sum, specialization in higher education is part of resident perception of the Piedmont urban complex for three of the institutions of higher learning. The fact that respondents could not characterize and differentiate among all such institutions in the area may have been due to a failure in interview technique or it may have been because the many institutions confuse the subjects and they therefore would not attempt to characterize any.

Perception of Area Government

It would seem reasonable that a group of closely-spaced urban nodes functioning as a single urban entity would benefit administratively from common government so that governmental services would be available equally to all. The Piedmont Triad Council of Governments (Chapter III) is a start in that direction although COG does not, as yet, govern. However, only ten per cent of the respondents were aware of the existence of COG (though none could correctly des-

cribe its structure and function), and only sixteen per cent of the
respondents thought a central, common government desirable for the
five-county area. Although a certain amount of local pride was
reflected in the responses, interviewees mostly thought a central
government was simply not necessary. "We're doing all right now,
why tamper with success?" was the general feeling. When pressed
to respond with specific reasons for opposition to a common government for the five county area, respondents listed cost, uncon-
trolled bureaucracy, and the alleged failure of northern metropolitan governments in that order. Although these answers may not
reflect reality, they very likely reflect a desire to maintain the
status quo and perhaps a desire to maintain the character of the
area as it is.

The resident of the region is aware of most of the services
received from local government and rates them at least adequate.
A surprising 42 per cent rated taxes as fair and commensurate with
services performed. Fifty-eight per cent rates taxes "high".
Forty-two per cent of the respondents believed taxes to be about
the same in comparable nearby cities or rural areas, thirty-eight
per cent believed them to be higher in other areas, ten per cent
believed them to be lower, and ten per cent had no opinion.

If it would be advantageous for the Piedmont urban complex
resident to live, administratively, under a common government, he
does not know it. Eighty-four per cent of the respondents do not
want a single government for the agglomeration and would oppose
such a move if suggested.

Shopping Habits

Burton suggests that the more affluent citizens of a dispersed
city have a reputation of going far afield to shop, usually to a

large metropolitan area.[1] The implication is that top-of-the-hier-
archy shopping is lacking or absent in the typical dispersed city.
A cursory survey reported in a prior publication seemed to substan-
tiate this statement.[2] This prior survey included only forty male
businessmen from Greensboro, all in higher income brackets. All
forty respondents took their wives or other family members out of
North Carolina at least three times a year for shopping and recreation. Further, a third of the respondents reported that their
wives made at least one additional trip per year for purposes of
shopping and recreation.

On the other hand, the **spatially** random sample of 350 respondents
in the area suggests a different picture than did the earlier survey.
Number of trips out-of-state averages slightly more than two per
year per respondent. This is not significantly different from the
slightly less than three trips per year per person in the U.S.,
overnight, to places one hundred or more miles away predicted for
the year 1976 by the National Planning Association.[3] Differences
become even less if the (almost) three predicted trips per year are
reduced by the 18 per cent of these trips that will be for business
purposes. Such trips were not recorded in the area resident sample.

The National Planning Commission predicts that 33 per cent of the
the non-business trips in 1976 will be for vacation, recreation, and
shopping purposes.[4] The resident survey indicates that 27 per cent
of the out-of-state trips are primarily for vacation and recreational
purposes, five per cent primarily for shopping, or a total of 32 per
cent devoted to these purposes. The National Planning Commission

---

[1]Burton, op. cit., p. 288.

[2]Hayes and Bennett, op. cit., pp. 12, 13.

[3]United States Department of Labor, Bureau of Labor Statistics,
National Planning Association, ORRC Study Report 23, op. cit.,
pp. 111-115.

[4]Ibid., p. 112.

predicts that 57 per cent of the non-business trips in 1976 will
be to visit friends or relatives. The resident survey indicates
that 55 per cent of the trips out-of-state are primarily for this
purpose. The National Planning Commission predicts that 10 per cent
of the non-business trips in 1976 will be devoted to personal or
family business. The resident survey indicates that 12 per cent
of the trips are for this purpose. Thus the number of trips and
primary trip purposes for the Piedmont urban complex resident do
not differ significantly from those predicted by the National
Planning Commission for the average United States citizen.

Secondary trip purpose differed from primary trip purpose
for the residents of the area. Of the 27 per cent whose primary
trip purpose was recreation, secondary purposes were as follows:
80 per cent for shopping, 10 per cent to visit relatives, 10 per
cent no secondary purpose. Of the 5 per cent whose primary purpose was shopping, 100 per cent offered recreation as the secondary
trip purpose. Of the 55 per cent whose primary trip purpose was
to visit friends or relatives, 80 per cent had a secondary purpose
of recreation, 10 per cent shopping, and 10 per cent none. Of the
12 per cent whose primary trip purpose was personal or family business, 50 per cent gave shopping, 40 per cent gave recreation and
10 per cent gave "none" as the secondary purpose. About one-third
of the total sample then offered shopping as a secondary trip
purpose.

Secondary trip purposes indicate that shopping facilities in
the study area are perceived as less than satisfactory. In fact,
residents expressed opinions concerning retail quality in the area.
Forty-five per cent of the respondents believed the last out-of-
state city visited had better shopping facilities than any node
within the urban complex. Of this forty-five per cent, sixty per
cent thought the selection of retail goods better, thirty-five

per cent thought prices lower, and five per cent thought merchandise
quality higher. When respondents were asked to rate retail shopping
facilities within the area on a five-point scale of "excellent",
"good", "adequate", "fair", and "poor", the distribution was bi-modal
at "fair" and "good, bracketing the median of "adequate".

This investigation indicates the resident of the study area
makes about the same number of trips per year out-of-state as does
the average person in the country and for about the same reasons.
Even though the resident perceives shopping facilities and recreational services as only adequate, this apparently does not act as a
significant push force generating an unusual number of trips (shopping or other) from the area. In other words, even though the resident rates shopping facilities only "adequate" here and "better"
elsewhere, he does not appear to act on this perception.

The Idea of City

Residents consider the Piedmont urban complex a very desirable
place to live (except for excessive automobile traffic friction,
inadequate garbage collection in some cases, and less then excellent shopping facilities) because of its "dispersed-city" character.
Residents can identify inter-nodal specialization in several functions, and they know that many people of the area live in one node,
shop in another, work in yet another, and so on. Does this "add up"
to the perception of the five county area as an urban entity? Is
the "idea of city" developed among its residents? The answer must,
as yet, be "no". Not only do most residents oppose the suggestion
of a single government for the urban complex, but 100 per cent of
the respondents answered "no" to the question: "In your opinion do
you think that Greensboro, Winston-Salem, and High Point form a
city unit?" and "yes" to the question: "Are they truly separate
cities?" The area resident does not yet confer "city" status on the
whole of the multi-nodal urban region.

Chapter VI

CONCLUSIONS AND IMPLICATIONS FOR PLANNING

Introduction

The study area consists of a group of urban nodes close to one another, separated by tracts of non-urban land with distances short enough for residents to choose one of several for important functions. Several cities within what might be called the Piedmont agglomeration are in the same population size-class, and there is no predominant urban node. Thus, in terms of morphology, the Piedmont urban complex appears to meet the criteria for "dispersed city" as defined in the literature. In fact, it is argued that if the title "dispersed city" could be conferred on the Piedmont urban complex on the basis of patterns and structure alone, it could be accepted without question. However, function and "perception of city" are additional published criteria for conferring or withholding the "dispersed city" title. The study area meets the first additional criterion only partially and the second not at all.

The residents of the study area do, indeed, often live in one node, shop in another, work in a third, seek higher education in another, buy goods that have been distributed from another node, and find recreational opportunities in yet another. Work, play, shopping, learning, and wholesaling are important functions and could combine to demonstrate that the study area meets the hypothesized criteria for defining a dispersed city in functional terms. Unfortunately, the degree of nodal functional interaction here falls somewhat short of the total interaction implicit in the literature. Still, central business trade area overlap is more than fifty per cent, implying considerable cross-commuting for shopping purposes. In fact, all respondents to the questionnaire reproduced in the appendix visited at least two central business districts, two-thirds of the respondents visited three central business districts, and 10 per cent of the respondents visited four central business districts with more or less regularity. Further questioning reveals that much cross-commuting is the result of perceived nodal retail specialization. It is unfortunate that nodal retail specialization is, very likely, more imagined than actual, since nodal specialization has been cited as a key characteristic of dispersed cities. Yet retail specialization is perceived, and a fair number of people act on the basis of this perception.

Cross-commuting for purposes of work is not as pronounced as is cross-commuting for purposes of shopping. Yet, there are some people who live in one node and work in another, others who live in a city and work in the country, and several who live in the county and work in one of the urban nodes.

In addition to the nodal interaction resulting from cross-commuting for purposes of shopping and work, nodal interaction results from cross-commuting for purposes of wholesale distribution, a desire for recreation,and a search for higher education.

In short, the concept of "dispersed city" fits the study reasonably well, but not perfectly. If the Piedmont urban complex is

compared to the dispersed city model implicit in the literature
on the basis of nodal interaction alone, it must be judged a partial, rather than a complete, dispersed city. Perhaps the Piedmont complex is still developing as a dispersed city. On the
other hand, Chauncy D. Harris suggests that:

> "Where the principal activity in the cities is not
> retail or wholesale trade or service activities for a
> surrounding local or regional area, often pre-dominantly
> rural, but in which the cities have manufactural activi-
> ties possibly related to a localized resource (as in coal-
> mining communities in Southern Illinois, in the Ruhr indus-
> trial district of Germany, or the Donbass in the Soviet
> Union) and with distant and often national markets for the
> products, one may have a belt of cities in close proximity
> though separated by rural tracts, with only modest rela-
> tionship to one another.
>
> The North Carolina Piedmont complex has some of these
> characteristics, with the three larger cities --Winston-
> Salem, High Point, and Greensboro-- all with very significant
> industrial employment, with industries partly related to
> local labor resources, partly rural or small town, but with
> national markets. The retail activities of each urban clus-
> ter may serve that cluster predominantly, though because of
> their proximity, residents of one have an option to work or
> shop in another (though the proportion doing so is apparently
> rather small).
>
> Would it be accurate to say that the area has some cha-
> racteristics of a dispersed city but that they are weak?
> Perhaps one of the features the area has in common with some
> other clusters of manufacturing cities, is the existence of
> a group of cities, in which no one clearly dominates the
> others, for the cities are not structured in a central-place
> hierarchy, but each of the industrial cities has separate
> ties to local resources and to national markets."[1]

Nevertheless, in spite of Professor Harris' perceptive comments,
it is suggested that the term dispersed city can be, at least tentatively, applied to the Piedmont complex. It is possible that
the area under study is at minimum, an incipient or developing
dispersed city, a conclusion indicated by the single market character of the area. The consumption of advertising media and
the orientation of marketing executives, both directed to serving
a single market, suggest this conclusion. Not only is the one-

---

[1]Chauncy D. Harris, personal communication (Chicago, Ill.,
Geography Department, The University of Chicago, February 9, 1975).

market concept already a greater part of the cognition and conceptual behavior of media and business executives than it is with the "man-on-the-street", but these executives can do much to further the concept with the average resident. It is suggested that the one-market, one-city idea becomes stronger each day because of the influence of the media and its supporters.

Dispersed cities have developed, or are developing, here and elsewhere, because of transport technology in the formative stage of urbanization and because the urban nodes, for one reason or another, grew at approximately the same rate. In the study area, the desire to serve the county resident's governmental needs limited the county size to one that permitted round-trip travel from periphery to center in one day by horse and wagon. This, of course, required each county seat to be in the approximate center of the county it served. The proclivity of southern manufacturers to locate manufacturing plants in smaller cities and towns as well as in the larger urban nodes kept the urban places growing at about the same rate until a specialty could be offered. Yet, in spite of the remarkable, if partial, dispersed city morphology and function in the area, each urban node remains a model miniature of a single-centered United States city with regard to population growth and distribution.

When for one reason or another, a cluster of towns has developed in a given area, new transport technology plus organizational capability and initiative may lead to a dispersed city system, involving a quantum jump from a pre-modern scale of interaction to a modern one. It is possible that the Piedmont agglomeration is on the brink of such a leap. If the nodes really assume specialized roles, if travel between them could be made inexpensive and convenient, and if the average citizen could come to regard himself as a resident of a single dispersed city, the area under study could be said to make the transition from a partial dispersed city to a comprehensive one.

This could be of great advantage to the area, for the typical
resident already likes living in a dispersed city environment and
likes the things about it that confer dispersed city character on
the place. However, the typical resident does not like the traffic situation, the shopping facilities, and the solid waste management efforts within the urban region. If implementation of long-
range plans could alleviate some of the problems associated with
these three factors, greater resident satisfaction would surely
result with or without complete dispersed city status. Planning
for the alleviation of the three problems can be a governmental
function without stretching the governmental role too far. The
obvious governmental unit for the job is the already existing
Piedmont Triad Council of Governments, as will be noted.

Transportation Planning

Transportation is the crux of dispersed city function. This
study reveals nothing concerning the chronology of transport
development in the study area; so it is not known when the Piedmont
agglomeration adopted some of the functional characteristics of a
dispersed city. It is possible that these characteristics developed
most rapidly between the years 1940 and 1970, however, for these
three decades marked a period of substantial increases in personal
income and of rapid improvement in transport technology throughout
the nation. It seems a reasonable a priori assumption that an
increase in personal income reduces the ratio of transportation
expenditure, and it seems equally reasonable that improved transport
technology reduces the time required for any given journey.
Nevertheless, even if inter- and intra-nodal transportation did
improve in the area over the thirty years prior to 1970, it is now
at the "squeaking point" as evidenced by the opinions of the
interview respondents. Transportation planning is, therefore, one

of the first orders of governmental concern, even though the area
resident does not blame his mobility problems on his elected and
appointed government officials.

Why ask for

High-speed rail rapid transit along existing thoroughfare *the*
...

medial strips or along existing rail lines has been suggested as a
reasonable method for improving accessibility among the various
urban nodes. Surely this suggestion has merit. The technology is
not beyond the knowledge or ability of the American people and cost,
though considerable, might be less in the long run than continued
dependence on the private automobile with attendant expenses of con-
stant highway construction, improvement, and maintenance, and given
almost certain higher costs of automobile fuel. A "park-n-ride"
feature would permit flexibility at the trip origin, and mini-
shuttle busses at the trip destination would permit routing to
various points within the destination node, although a "dial-a-bus"
feature would be necessary in order for the transit rider to return
to the point-of-destination for the trip home.

The relative magnetism of various functions within each node
could easily be compiled through reference to traffic O-D surveys.
The desire-lines produced by such surveys could be converted to the
desire to journey to a destination by public conveyance as well as
by private auto. The problem is behavioral, not technological.
Such a transport system would transfer a certain amount of responsibility and direct cost from the private to the public sector. The
question is whether elected officials, publicly dedicated to reduc-
ing taxes, could, in fact, raise them and be re-elected even though
a cost increase in the public sector would be offset by a decrease
in the cost of private transportation. Would the residents of the
"Piedmont Dispersed City" be willing to give up the flexibility of
automobile travel between urban nodes for a system offering less
flexibility even though such a system would help protect dispersed

Confusion here

city function for future generations? The answer is not known, but a shift in transport media emphasis can only be accomplished through governmental action.

Improved intra-nodal travel would also improve inter-nodal travel. Automobile congestion within the various urban nodes of a dispersed city reduces travel convenience and increases travel time from any origin to any destination. Bicycle rights-of-way along collector streets and thoroughfares might not put every citizen on a cycle, but at least those willing to pedal would not fear for their lives on the now busy streets and roads. Sidewalks, in all nodes, connecting residential districts with higher use functions might induce some residents to walk a few blocks instead of firing up the car. Present bus routing, flexibility, and image could also be improved within the various nodes of the urban region. Such improvements would surely capture more riders, for an unpublished study by the author for the Greensboro Planning Department in 1967 indicated that the two major reasons that less than ten per cent of the trips within Greensboro were by bus were routing and prejudice; that is, bus routes did not serve the Greensboro traveler's trip purpose adequately, or, if they did, busses were considered low-class vehicles suitable for conveying domestic workers to the homes of the affluent.

In any case, intra-nodal transport changes, as is the case with inter-nodal transport changes, can only be accomplished through governmental influence or intervention. Perhaps the best thing that could happen to improve the mobility of the dispersed city resident would be a chronic energy shortage. Such a continuing crisis might move the governmental bureaucracy to action quicker than any other single event.

Retail Planning

The local governmental unit concerned with the urban region --
the Piedmont Triad Council of Governments -- neither governs nor is
concerned primarily with the "Piedmont Triad." The triad is usually
defined as the urban nodes of Greensboro, Winston-Salem, and High
Point and the non-urban land between the three nodes.

COG is concerned with eleven counties, the five comprising the
Piedmont urban complex plus six additional, predominantly rural,
counties to the north and west of the five. Since the problems and,
thus, the objectives of the five counties vs. the six are different--
a desire for the control vs. a desire for growth[1], conflict might
be reduced and plans might become more readily implementable if
the COG official concern was to exclude the six rural counties
located outside the boundaries of the Piedmont agglomeration. This
is true especially because the COG depends on the support of the
individual local governmental units for long-range plan acceptance,
and the conflict of objectives makes agreement among the eleven
counties difficult.

If this reduction in scope were accomplished, perhaps the
Council of Governments could persuade the United States Bureau of
the Census to accept the five-county area as a single Standard
Metropolitan Statistical Area. An SMSA of nearly eight hundred
thousand people would be an important metropolitan unit within
the country. Its creation might stimulate public attention in the
whole area rather than on the individual units.

Such an SMSA could support major retail outlets instead of the
now rather small stores operating within each urban node. Nodal
specialization would, of course, be necessary since all nodes
would not support major retailing. However, specialization would

---

[1]Hayes, Bennett, Sowers, Regional Development Guide, op. cit.,
pp. 1-10.

enhance the dispersed city character of the urban region, a characteristic that the residents find quite appealing about the place.
Perhaps major retail outlets are absent from the urban region because chain-store management, operating from a distance, does not perceive the dispersed city character and function of the region, but this is conjecture. Moreover, it is possible that retail management realistically perceives that most retail purchases focus on the closest node, on a few small discrete cities rather than on one sizable dispersed city. If this is the problem, it is behavioral and can be changed by governmental action and by diffusion and acceptance of new images and perceptions. It is suggested that the first step in the solicitation of major retail outlets is the recognition of the Piedmont Dispersed City, if it may now be called that, as a single Standard Metropolitan Statistical Area, providing a large market. Such recognition might also help foster the "idea of the city" as a necessary further step in planning for the development of the area.

Solid Waste Management Planning

All respondents who depend on private contractors for solid waste collection, or who must dispose of solid waste themselves, are dissatisfied with the lack of service and all blame the local governmental unit for not providing the service. Solutions to the problems of solid waste management for the five-county area are surely available even though a search of the COG publications does not provide them. Perhaps solid waste collection and disposal are best handled at the local rather than regional level. Perhaps each county should assume responsibility for these services or perhaps it is the proper responsibility of the Piedmont Triad Council of Governments. At any rate, it is a sensitive point to all residents of the five-county area, especially to those who

believe they are not now adequately served. Solid waste management
is a point of entry for governmental public relations.

Population Planning

If population growth and distribution for the Piedmont Dispersed
City is to be planned, the first task facing the Piedmont Triad
Council of Governments is to take its program directly to the people.
Residents of the urban region fear growth because they fear it will
obliterate the present form of urban agglomeration amid encompassing
rural stretches. Residents deplore the possibility of a situation
wherein the nodes of the urban region coalesce to form a continuous
strip of urbanism. If this view is correct, a majority of the residents of the Piedmont Dispersed City favor the COG population growth
and distribution plan over the plan proposed by the Greensboro
Chamber of Commerce. The COG plan to limit population growth by
limiting industrial in-migration and channeling its location would
seem to be sound, though there is no record of its having been successful elsewhere. There is no reason to believe the plan is not
implementable if the support of the people can be obtained, since
local governmental support would soon follow popular support. At
any rate, it is not too soon to study the situation in order to cate-
gorize desirable potential manufacturing in-migrants by standards
of pollution potential, energy use, employment potential, wage rates,
and of course, by location factors and location orientation so that
solicitation efforts are efficiently and realistically applied.

Control of population distribution for the urban region should
prove no more difficult than control of population growth, though
the results of neither are well documented as a pragmatic project.
The combination of constraints proposed by the COG --area-wide
zoning, city boundary control, utility and service control,

and green belts -- might effectively promote any reasonable distribution of the population upon which local governmental units can
agree. Desirable uses of the land within the Piedmont Dispersed
City for residential purposes and for commercial, industrial,and
institutional purposes may be a more difficult question to settle.
At any rate, it is not too soon to develop a five-county comprehensive land-use plan.

The Idea of the City

An important task facing the Piedmont Triad Council of Governments, whether a consortium of eleven or five counties, is the
development of a program resulting in public recognition of COG
as a major governmental entity. As yet, residents of the Piedmont
Dispersed City barely know of its existence and know nothing of
its structure and functions. Even though the area resident knows
that some of his neighbors depart for work, shopping, play, education, and so on, in different directions, bound for different
destinations, and even though the area resident perceives nodal
specialization, he does not recognize the five-county urban agglomeration as a single city in functional terms. It is suggested
that public recognition of the COG governmental unit is a first
step in promoting the "idea of the city". If the dispersed city
resident can look to a single governmental unit for the provision
of services and for the maintenance of public welfare, a step
toward accepting the idea of city surely would result.

The idea of the city could be promoted by the communications
media serving the area. Inasmuch as the executives and managers
of the radio, newspaper, and television companies already perceive
the five-county area as a single market, it is surely only a small
step to applying the idea of the city to the five-county urban
agglomeration. If each unit of each medium promoted the concept

of a single city, if only by slogans such as "serving the Piedmont Dispersed City," the idea might eventually be accepted. Retail merchants also might promote the idea of the city through advertising their speciality dispersed city-wide. Inasmuch as most area residents believe that inter-nodal specialization already exists, this gambit would reinforce the perception and result in furthering the concept of a functional unit for the five-county agglomeration.

The idea of the city also might be promoted by the formation of a single chamber of commerce to serve the area, by a single public school administration, a single branch of the University of North Carolina, and so on. If a city is an idea, a percept, a belief, as well as a physical entity, it is essential that the typical resident of the area begins to think "city" for what he now perceives as an agglomeration of towns, vaguely linked, and to modify a set of values deeply rooted in a nostalgia for a past which has no future in a rapidly changing metropolitan society.

Appendix 1

QUESTIONNAIRE

The questionnaire following was prepared initially and pre-
tested in the fall of 1970. The pre-test was conducted by the
four college student interviewers who would obtain the final interviews, and the pre-test served the dual purpose of interviewer
training and questionnaire refinement. Upon completion of pre-
test and interviewer conferences concerning pre-test, the questionnaire was revised to its present form. Revisions included
eliminating or changing a few words that appeared controversial in
pre-test and changing the order of the questions. Also, since
the questionnaire requires both factual and non-factual (opinion)
responses it was considered important to eliminate "leading"
questions except where this was deliberate.

The order of the questionnaire was designed to get the subjects into the spirit of being interviewed and to get them "talking".
Therefore, the questionnaire started with factual or "easy-to-
answer " questions and proceeded to opinion or "hard-to-answer"
questions. This order was determined during pre-test. The questionnaire itself can be completed in about a half-hour, but interviewers were instructed to converse with subjects in order to obtain in-depth opinions. This effort consumed roughly an extra
half-hour per respondent. Some of the conclusions reported in
Chapter V are based on these in-depth opinions.

There are certain specific items in the questionnaire that require comment. For example, the title "Applied Academic Research" is the name of a consulting firm with which the author is associated. Question Number 49 concerning traffic congestion is obviously leading but was so intended. It was asked after the respondent had an opportunity to voice general area dislikes. Question 49 was included on the a priori assumption that traffic congestion is a problem within the study area. Questions 52 and 53 concerning COG are also leading but, again, were asked after a general question concerning government. Questions 64, 65, 66, and 67 concerning garbage collection were included also on the a priori assumption of problems within the study area. Sex, race, age, and income were judged by the interviewer and were included only to insure a demographic cross-section; that is, tabulation was not based on these factors.

Interviews for the spatially random sample of the entire area were obtained during March, April, and May of 1971. Spatial randomness was obtained through the use of map grids with 1,000-square-foot cells selected with a table of random numbers. Interviewers were instructed to approach the residence closest to the center of the appropriate cell, but if a successful interview could not be obtained, then to approach the next closest and so on until an interview could or could not be obtained. If no interview could be obtained in a particular cell, it was discarded and another cell was selected at random. Two hundred successful interviews were obtained in this fashion.

Interviews for the spatially random sample of the six major urban nodes were obtained in March, April, and May of 1973 by two of the four college students who had worked on the prior survey. All techniques and questions were the same as in the prior survey.

No basic opinion or factual answer changes were detected in the two surverys, and the only significant demographic change was the inclusion of considerably more Negro respondents in the second survey. One hundred and fifty successful interviews were obtained; thirty each from Greensboro and Winston-Salem, twenty-five each from Asheboro and High Point, and twenty each from Burlington and Lexington. The primary reason for two surveys was the desirability of using the entire area survey to synthesize trade areas and labor sheds in addition to the facts and opinions obtained from respondents. A strong secondary reason, however, was that a two-year space between surveys spread out the cost which was not inconsiderable.

No unusual problems were reported by interviewers regarding finding correct houses or obtaining successful interviews, although a few cells were discarded either because they contained no houses (in rural areas) or because no one was at home. No one refused to answer the questions. The interviews were considered success-ful.

I am conducting a survey for Applied Academic Research concerning
your attitudes and opinions of this area. It will take about half
an hour to answer the questions, but we will appreciate it very
much if you will give us these opinions by answering a prepared
list of questions. Thank you.

1) Where do you buy groceries most often?

2) Where do you buy clothing most often?

3) Where does your husband (wife) buy clothing most often?

4) Where do you usually buy appliances (or last appliance)?

5) Automobile?

6) Which shopping center do you use most often?

7) Does (shopping center) offer better quality?   values?
    prices?   unique products?
    explain -

8) Do you like a particular store in (shopping center)?
    What store?
    Why?

9) Are there any other reasons you shop at (shopping center)?

10) Why did you visit (shopping center) the very first time?
    explain -

11) Do you ever read ads in newspapers or hear ads on radio or TV that relate to (shopping center)?

What ad?

Product or store?

Media?

12) What is your favorite TV ad?

13) Radio?

14) Newspaper?

15) Where do you work?

16) Where does your husband (wife) work?

17) When is the last time you shopped downtown?

City?

18) How often do you shop downtown?

19) Do you ever shop at a downtown other than the one you just mentioned?

Where?

How often?

20) Do you ever shop at a third downtown other than the ones you just mentioned?

Where?

How often?

21) Do you ever shop at a fourth downtown other than the ones you just mentioned?

Where?

How often?

22) Do you think downtown offers good quality?   values?
prices?   unique products?
explain-

23) When is the last time you visited a nearby city?
City?

24) What was the purpose of your visit?

25) What nearby cities do you visit fairly often?

]6) How often?

27) What are your usual reasons for visiting nearby cities?

28) If you owned a company and wanted to buy employee insurance
where would you go to buy it?
Why?

29) Company advertising?
Why?

30) Medical insurance?
Why?

31) Where is your own doctor located?

32) Furniture for your company offices?

33) Product photography?

34) What, in your opinion, is the best shopping center in the Triad?

Why?

35) Which city has the best downtown?

Why?

36) Which city has the next best downtown?

Why?

37) Which city has the best government?

Why?

38) Police force?

Why?

39) Garbage collection?

Why?

40) Do you believe there are any real differences in the cities of the Triad or do you think they are all much the same?

explain-

41) Do you believe different cities in the Triad specialize in different goods and/or services?

What?

explain-

42) If you had your choice where would you live in the Triad area?

43) Anywhere in the country?

Why?

explain-

44) How long have you lived in this area?

45) Where did you live before you moved here?

46) What do you like about living here?

47) What do you dislike about living here?

48) Is traffic congestion a problem to you?

49) Do you think the Triad would be better off under a single government or do you prefer things the way they are?

50) What governmental units influence your life?

51) What is the Piedmont Triad Council of Governments?

52) Tell me what you know about it (COG).

53) Do you think your local taxes are too high for services rendered or about right?

54) In your opinion are taxes lower or higher in nearby cities?

55) In the county?   (city?)

56) Would you move to the county (city) if you could?
    Why?

57) What do you like about the present city (county) administration?

58) What do you dislike about the present city (county) administration?

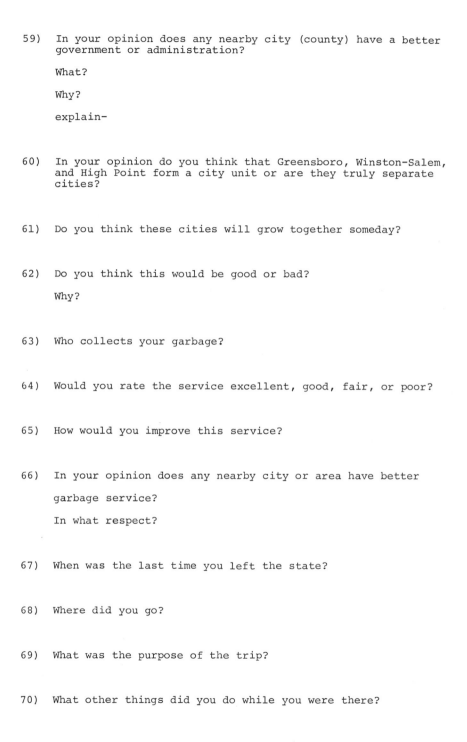

143

59) In your opinion does any nearby city (county) have a better government or administration?

What?

Why?

explain-

60) In your opinion do you think that Greensboro, Winston-Salem, and High Point form a city unit or are they truly separate cities?

61) Do you think these cities will grow together someday?

62) Do you think this would be good or bad?

Why?

63) Who collects your garbage?

64) Would you rate the service excellent, good, fair, or poor?

65) How would you improve this service?

66) In your opinion does any nearby city or area have better garbage service?

In what respect?

67) When was the last time you left the state?

68) Where did you go?

69) What was the purpose of the trip?

70) What other things did you do while you were there?

71) About how many times a year do you leave the state?

72) Where do you usually go?
    Why?

73) What did you buy the last time you left the state?

74) Does (city) have better shopping facilities than here?

75) In what way?

76) Would you prefer to live in (city)?
    Why?

77) Tell me your general feeling about living in this area.
    explain-

78) Can you rate shopping facilities in the area as:
    excellent?
    good?
    adequate?
    fair?
    poor?

79) Where is Wake Forest University located?

80) Where is UNC/G located?

81) Where is Guilford College located?

82) Can you rate the instructional quality of the three?

Wake Forest   Excellent, 2, 3, 4, Poor

UNC/G   Excellent, 2, 3, 4, Poor

Guilford   Excellent, 2, 3, 4, Poor

83) Tell me what you know about:

Wake Forest

UNC/G

Guilford-

84) What is the most important product manufactured in:

Winston-Salem-

Greensboro-

High Point-

Burlington-

Lexington-

85) What is the second most important product manufactured in:

Winston-Salem-

Greensboro-

High Point-

Burlington-

Lexington-

MALE   FEMALE   WHITE   OTHER(what)

AGE: 20-30;   30-40;   40-50;   50-60;   60 plus

HOUSE: size:

        age: new, 10;   20-40;   40 plus

        condition: excellent, good, fair, poor, deteriorating

LOCATION:

INTERVIEWER AND OTHER COMMENTS:

147

Appendix 2

PUBLICATIONS OF THE
PIEDMONT TRIAD COUNCIL OF GOVERNMENTS*

SOLID WASTE MANAGEMENT
Regional Solid Waste Disposal Study: Phase I - April, 1972 - 34 pp.
Regional Solid Waste Management Study: Phase II - April, 1973- 94 pp.

TRANPORTATION
The 1972 National Transportation Needs Study: Summary Report -
    August 1971, 15 pp.
The 1972 National Transportation Needs Study: Piedmont Triad Region-
    2 Volumes - Vol. I (narrative), 192 pp. + appendices; Vol. II
    (forms), 258 pp.
The 1972 National Transportation Needs Study: High Point Urban Area-
    August 1971 - 124 pp. + appendices
The 1972 National Transportation Needs Study: Greensboro Urban
    Area - August 1971 - 128 pp. + appendices
The 1972 National Transportation Needs Study: Winston-Salem Urban
    Area - August 1971 - 125 pp. + appendices
The 1974 National Transportation Study: Narrative for the Units
    of Government in Region G - August 1973 - 87 pp. + appendices

PROJECT REVIEW
Report on the Project Notification and Review System, State Multi-
    County Region G, April 1, 1971 - March 31, 1972   May 1972 -
    10 pp.
Report on the Piedmont Triad Council of Governments Project
    Notification and Review System, April 1, 1972 - March 31, 1972 -
    May 1973 - 6 pp.
A-95 Project Notification and Review System, Piedmont Triad Council
    of Governments - 1972 - Fold-out brochure

REGIONAL PLANNING AND DEVELOPMENT
Population, Economy, and Land Use Study, Regional Development
    Guide: Phase I - Summary Report, May 1971 - 18 pp.
Population, Economy and Land Use, Regional Development Guide:
    Phase I - May 1971 - 105 pp.
Concept Guide for Regional Development, Piedmont Triad Region,
    Region Development Guide Program: Phase II - May 1972 -
    83 pp.
Regional Development Guide - Nov. 1972 - Fold-out brochure
Report on Regional Development Guide - Phase III, 1972-73 -
    May 1973 - 10 pp.

DEVELOPMENT STANDARDS AND CODES
A development Standards and Codes Study - January 1971 - 92 pp.

*Note: This bibliographical appendix is reproduced exactly as
received from the Piedmont Triad Council of Governments.

WATER SUPPLY AND WASTEWATER DISPOSAL

Regional Water Supply and Wastewater Disposal Study, Inventory and
Problem Delineation, Phase I Report - June 1971 - 54 pp.
Regional Water Supply and Wastewater Disposal Study, Phase II,
Interim Report - October 1972 - 209 pp.

COMMUNITY/CITIZEN PARTICIPATION (INFORMATION)

Citizen Participation in the Piedmont Triad Council of Governments
1971-1972 - May 1972 - 6 pp.
Community Participation in the Piedmont Triad Council of Governments,
1972-1973 - May 1973 - 7 pp.
Report on Piedmont Triad Council of Governmental Staff Contracts
with Governmental Agencies, January 1 - Mar 31, 1973 - May 1973-
4 pp.
The Role and Purpose of the Piedmont Triad Council of Governments,
1971-72 - July 1971 - 4 pp.
A look at PTCOG, January 1973 - Fold-out brochure with tabs
Report on First Annual Piedmont Council of Governments Delegate
Retreat, April 1973 - May 1973 - 3 pp. with attachments
An Analysis of Neighborhood Planning Center (Model Cities Project)
Dec. 1971 - 81 pp.
Supplementary Report - Neighborhood Planning Center (Model Cities
Project) - Dec. 1971 - 12 pp.
COG LOG - Newsletter of Piedmont Triad Council of Governments-
published bi-monthly

HOUSING

COG Initial Housing Element in the Piedmont Region G, North Carolina -
April 1970 - 32 pp.
COG Housing Element II in North Carolina Region G - April 1971 -
25 pp.
Report on the Housing Program of the Piedmont Triad Council of
Governments 1971-72 - May 1972 - 8 pp.
Report on the Housing Program of the Piedmont Triad Council of
Governments, 1972-1973 - May 1973 - 15 pp.
Land Bank Handbook (Advance Acquisition of Sites for Low and
Moderate Income Housing) - January 1972 - 221 pp.

OPEN-SPACE/RECREATION

Regional Open Space and Recreation Study: Phase I - May 1972 -
96 pp.
Regional Open Space Study: Phase II-April 1973 - 88 pp.
Piedmont Triad Park - 1968 - 4 pp.

CRIMINAL JUSTICE

Program of Action for Fiscal Year 1972 - Piedmont Triad Criminal
Justice Planning Unit of the Piedmont Triad Council of Governments - February 1971 - 190 pp. + appendices

ENVIRONMENTAL AFFAIRS

Proceedings of the Regional Environmental Workshop - Nov. 1971 -
24 pp.

GENERAL
Revised Charter of the Piedmont Triad Council of Governments - May 17,
      1972 - 4 pp.
By-Laws of the Piedmont Triad Council of Governments - Feb. 15, 1972-
      4 pp.
A Program of Work for the Piedmont Triad Council of Governments -
      January 1970 - 51 pp.
Proceedings - Regionalism in the 70's Conference - Nov. 1971 - 75 pp.
Description of Proposed 1972-73 Program of Work Elements - February
      1972 - 16 pp.
Issues Affecting PTCOG's Future Role - April 1973 - 3 pp.

MANPOWER
Area G Comprehensive Manpower Plan, Fiscal Year 1973 - Part A- March
      1972 - 19 pp. + appendices
Region G Manpower Plan, Fiscal Year 1974, May 1973 - 46 pp.

COMPREHENSIVE HEALTH
Health Goals Statement, Piedmont Triad Regional Planning Council of
      the Piedmont Triad Council of Governments - September 1972 -
      33 pp.

LIBRARIES
Regional Library Services Study - November 1973 - 107 pp.

BIBLIOGRAPHY

Journal Articles

Applebaum, William and Spears, Richard. "Controlled Experimentation
    in Marketing Research," The Journal of Marketing (Jan. 1950),
    pp. 505-517.

Applebaum, William and Spears, Richard. "How to Measure A Trading
    Area," Chain Store Age, Jan. 1951, revised brochure, no pp.
    nos.

Beimfohr, O.W. "Some Factors in the Industrial Potential of Southern
    Illinois," Transactions of the Illinois State Academy of Science
    46 (1953): 97-103.

Braschler, Curtis H. "Importance of Manufacturing in Area Economic
    Growth, " Land Economics 47 (Feb. 1971): 109-111.

Burton, Ian. "A Restatement of the Dispersed City Hypothesis,"
    Annals of the Association of American Geographers 53 (1963):
    285-289.

Burton, Ian. "Retail Trade in a Dispersed City," Transactions of
    the Illinois State Academy of Science 52 (1959): 145-150.

Ginsburg, Norton S. "The Dispersed Metropolis: The Case of Okayama,"
    The Toshi Mondai, Tokyo 52 (1959): 145-150.

Hartshorne, Turman. "The Spatial Structure of Socioeconomic Develop-
    ment in the Southeast," Geographical Review 61, No. 2 (April
    1971): 46-51.

Hayes, Charles R.and Schul, Norman. "Some Characteristics of Shop-
    ping Centers," Professional Geographer XVII, No. 6 (Nov. 1965):
    11-14.

Hayes, Charles R. and Schul, Norman. "Why Do Manufacturers Locate
    in the Southern Piedmont?," Land Economics XLIV, No. 1 (Feb.1968):
    117-121.

Lonsdale, Richard E. and Browning, Clyde E. "Rural-Urban Locational
    Preferences of Southern Manufacturers," Annals of the Association
    of American Geographers 61, No. 2 (June 1971): 255-268.

Terjung, W.H. "Physiologic,Climates of the Conterminous United
    States: A Bioclimatic Classification Based on Man," Annals of
    the Association of American Geographers 56, No. 1 (March 1966):
    141-179.

Vance, James E. "Labor-Shed Employment Field,and Dynamic Analysis
    in Urban Geography," Economic Geography 36, No. 3 (July 1969):
    189-220.

Books

Arnett, Ethel S. Greensboro, North Carolina: The County Seat of Guilford. Chapel Hill, N.C.: University of North Carolina Press, 1955.

Berry, B. J. L. Geography of Market Centers and Retail Distribution. Englewood Cliffs, N.J.: Prentice-Hall, 1967.

Brookfield, H.C. "Environmental Perception," in Progress in Geography. London: Arnold Press, Vol. I, 1969.

DeVyer, Frank T. et al. Labor in the Industrial South. Charlottesville, Va.: Michie Co., 1930.

Fuchs, Victor R. Changes in the Location of Manufacturing in the United States Since 1929. New Haven, Conn.: Yale University Press, 1962.

Hoover, Calvin B. and Rutchford, B. U. Economic Resources and Policies of the South. New York: Macmillan Co., 1951.

McLaughlin, Glenn E. and Robock, Stefan. Why Industry Moves South. Kingsport, Tenn.: Kingsport Press, 1949.

Neisser, Ulric. Cognitive Psychology. New York: Meredith Co., 1967.

Prohansky, H.M., Ittleson, W.H. and Rivlin, L. G. "The Influence of the Physical Environment on Behavior: Some Basic Assumptions," in Idem, Environmental Psychology. New York: Holt, Rinehart, and Winston, 1970.

Reichtel, Levin T. The Moravians in North Carolina. Baltimore: Genealogical Publishing Co., 1968.

Roundthaler, Edward. The Memoraphilia of Fifty Years 1877-1929. Raleigh, N.C.: Edwards and Broughton Co., 1928.

Sharp, Bill. A New Geography of North Carolina. Raleigh, N.C.: Sharp Company, 1958.

Stem, Thad Jr. and Butler, Alan. Senator Sam Ervin's Best Stories. Durham, N.C.: Moore Publishing Co., 1973.

Stockton, Sallie W. The History of Guilford County, North Carolina. Nashville, Tenn.: Gaut-Ogden Co., 1902.

Trewartha, Glenn T. A Geography of Population: World Patterns. New York: John Wiley and Sons, 1969.

Trotter, William. "Four Hour Thunder," Red Clay Reader, No. 6. Charlotte, N.C.: Southern Review, 1969.

Reports, Newspaper Articles, and Government Documents

Asheboro Chamber of Commerce. Asheboro, North Carolina, Basic Industrial Data. Asheboro, N. C. (undated).

Bennett, D. Gordon and Hayes, Charles R. Downtown Shopper Characterstics and Media Coverage in North Carolina. Raleigh, N. C.: North Carolina Department of Administration, State Planning Division, Report No. 112.08, May 1970.

Berry, B. J. L., Mayer, H.M., et al. Comparative Studies of Central Place Systems. Report to : Geography Branch, Office of Naval Research by Department of Geography. Chicago, University of Chicago, Department of Geography, 1962.

Bolden, Donald E. Alamance Battleground Bi-Centennial Commemorative Souvenir Program. Burlington, N.C., May 1971.

Bucher, George. Investigation of Local Resources for the Social Studies in Alamance County, A looseleaf Collection of Source Materials for Local History and Social Problems. Burlington, N.C., unpublished 1939-40.

Burgess, Fred. Randolph County: Economic and Social [A Laboratory Study of the University of North Carolina]. Chapel Hill, N.C., Department of Rural Social Economics, 1924.

City of Greensboro. Department of Planning. Land Use Plan. Greensboro, N.C., July 1967.

Fantus Company, Industrial Location Appraisal: Guilford County. New York, 1967.

Forbis, Charles O. and Lynch, Parker. Population: Guilford County, N.C. Greensboro, N.C. 1968.

Greensboro Chamber of Commerce, Research Division. Wholesaling in Greensboro. Greensboro, N.C. 1968.

Hammer, Greene, Siler Assoc. Downtown Burlington: An Analysis of its Economic Potential. Washington, D.C. July 1967.

Hammer, Greene, Siler Assoc. Forsyth County's Economic Prospect. Washington, D.C. April 1970.

Hammer, Greene, Siler Assoc. The Guilford County Economy. Washington, D.C. 1966.

Hayes, Charles R. and Bennett, D. Gordon. Factors of Spatial Interaction in North Carolina. Raleigh, N.C., North Carolina Department of Administration, State Planning Task Force, Report No. 64.13, April 1969.

Hayes, Charles R. and Bennett, D. Gordon. "Governmental Attitudes on a Stationary Population for the Piedmont Dispersed City", paper read to the North Carolina Population Center, Chapel Hill N.C. Feb. 1973.

Hayes, Charles R., Bennett, D. Gordon, and Sowers, Linda. Phase I Regional Development Guide, Piedmont Triad Council of Governments. Greensboro, N.C., 1971.

Hayes, Charles R. "Greensboro's Downtown Trade Area." Greensboro, N.C. 12 July 1967. Planning Notes, Planning Department.

Hayes, Charles R. Greensboro Recreation Survey. Greensboro, N.C. 1972, Greensboro Chamber of Commerce.

Hayes, Charles R. and Schul, Norman. Greensboro Retail Core Analysis. Greensboro, N.C. March 1965, Greensboro Planning Department.

Hayes, Charles R. and Schul, Norman. Greensboro Shopping Center Trade Areas. Greensboro, N.C., Greensboro Planning Department, May 1964.

Hayes, Charles R. Community Development. Greensboro, N.C., Gateways, 1973.

Lexington Chamber of Commerce, Lexington, North Carolina. Lexington, N.C., 1971.

Macintosh, Mary L. "History of Elon College," Diamond Jubilee, 1893-1968, Elon College (1968).

Marks, Robert. Greensboro Daily News, Greensboro, N.C. (May 29, 1971).

National Planning Association, Committee on the South. New Industry Comes to the South. Washington, D.C. (1949).

Neilson, Robert W. History of Government, City of Winston-Salem, N.C. Winston-Salem, N.C., 1966.

Paul, C.A. Greensboro Daily News, Greensboro, N.C. (May 29, 1971).

Philbrick, Allen K., Analysis of the Geographical Patterns of Gross Land Use and Changes in Numbers of Structures in Relation to Major Highways in the Lower Half of the Lower Peninsula of Michigan. East Lansing, Michigan State University, 1961.

Register, Robert E. Greensboro Daily News, Greensboro, N.C.(May 29, 1971).

Saarinen, Thomas F. "Perception of Environment," Association of American Geographers, Commission on College Geography, Resource Paper No. 5 (1969).

United States Department of Agriculture, Technical Bulletin 1210 (Washington, D. C., 1962).

United States Bureau of the Census, Census of Population (1899).

United States Bureau of the Census, Census of Population, Washington, D. C. (1930).

United States Bureau of the Census, Census of Population, Washington, D. C. (1970).

United States Bureau of the Census, Statistical Atlas of the United States, Washington, D. C. (1914).

United States Bureau of Labor Statistics, National Planning Association ORRRC Study Report 23, Washington, D. C. (1962).

Whitaker, Walter. "E.M. Holt and the Cotton Mill", Centennial
    History of Alamance County, Burlington, N. C., Chamber of Com-
    merce (1949).

Winston-Salem Chamber of Commerce. A Half Century of Progress 1885-
    1935, Winston-Salem, N. C. (Sept. 1935).

INTERVIEWS, CONVERSATIONS PERTINENT TO THE STUDY

Area Businessmen

William Alexander - Sales Manager WFMY-TV
John Bonitz - President Bonitz Insulation Co.
Robert Dabbs - President Dabbs Furniture Company
Leon Kiser - Head, Shipping Department, Sears Roebuck Catalog
    Warehouse.
Ian McBride - Director Community Relations WFMY -TV
John Morgan - Vice President Starmount Land Development Company
Thomas Pickard - Vice President Weaver Construction Company
Wm. Saunders - Advertising Manager - News/Record Newspaper
Wm. Snyder - Editor News/Record Newspaper
Dave Wright - News Director WFMY-TV
David Zauber -President Greensboro Board of Realtors
Gary Pannell - Location Specialist, Wachovia Bank

City of Greensboro

The Honorable Jim Melvin - Mayor
The Honorable Mack Arnold - Councilman
Charles Mortimore - Planning Director
Robert Barclay - Director Urban Renewal
James Isler - Director Community Development
Steve Davenport - Asst. Planning Director
Arthur Davis - Planner
William Chambliss - Planner

Greensboro Chamber of Commerce

William Little - Executive Director
John Parramore - Director Industrial Division
Thomas Routh - Director Urban Affairs

North Carolina State Officials

The Honorable McNeill Smith- State Senator
Ronald Scott - State Planning Officer
John Booth - Director Division of COG's

Piedmont Triad Council of Governments

Lindsay Cox - Executive Director
Wm. Colonna - Asst. Director
Chrys Constable - Planner

THE UNIVERSITY OF CHICAGO
DEPARTMENT OF GEOGRAPHY
RESEARCH PAPERS (Lithographed, 6×9 Inches)

(Available from Department of Geography, The University of Chicago, 5828 S. University Ave., Chicago, Illinois 60637. Price: $6.00 each; by series subscription, $5.00 each.)

106. SAARINEN, THOMAS F. *Perception of the Drought Hazard on the Great Plains* 1966. 183 pp.

107. SOLZMAN, DAVID M. *Waterway Industrial Sites: A Chicago Case Study* 1967. 138 pp.

108. KASPERSON, ROGER E. *The Dodecanese: Diversity and Unity in Island Politics* 1967. 184 pp.

109. LOWENTHAL, DAVID, et al. *Environmental Perception and Behavior* 1967. 88 pp.

110. REED, WALLACE E. *Areal Interaction in India: Commodity Flows of the Bengal-Bihar Industrial Area* 1967. 210 pp.

112. BOURNE, LARRY S. *Private Redevelopment of the Central City: Spatial Processes of Structural Change in the City of Toronto* 1967. 199 pp.

113. BRUSH, JOHN E., and GAUTHIER, HOWARD L., JR. *Service Centers and Consumer Trips: Studies on the Philadelphia Metropolitan Fringe* 1968. 182 pp.

114. CLARKSON, JAMES D. *The Cultural Ecology of a Chinese Village: Cameron Highlands, Malaysia* 1968. 174 pp.

115. BURTON, IAN; KATES, ROBERT W.; and SNEAD, RODMAN E. *The Human Ecology of Coastal Flood Hazard in Megalopolis* 1968. 196 pp.

117. WONG, SHUE TUCK. *Perception of Choice and Factors Affecting Industrial Water Supply Decisions in Northeastern Illinois* 1968. 96 pp.

118. JOHNSON, DOUGLAS L. *The Nature of Nomadism* 1969. 200 pp.

119. DIENES, LESLIE. *Locational Factors and Locational Developments in the Soviet Chemical Industry* 1969. 285 pp.

120. MIHELIC, DUSAN. *The Political Element in the Port Geography of Trieste* 1969. 104 pp.

121. BAUMANN, DUANE. *The Recreational Use of Domestic Water Supply Reservoirs: Perception and Choice* 1969. 125 pp.

122. LIND, AULIS O. *Coastal Landforms of Cat Island, Bahamas: A Study of Holocene Accretionary Topography and Sea-Level Change* 1969. 156 pp.

123. WHITNEY, JOSEPH. *China: Area, Administration and Nation Building* 1970. 198 pp.

124. EARICKSON, ROBERT. *The Spatial Behavior of Hospital Patients: A Behavioral Approach to Spatial Interaction in Metropolitan Chicago* 1970. 198 pp.

125. DAY, JOHN C. *Managing the Lower Rio Grande: An Experience in International River Development* 1970. 277 pp.

126. MAC IVER, IAN. *Urban Water Supply Alternatives: Perception and Choice in the Grand Basin, Ontario* 1970. 178 pp.

127. GOHEEN, PETER G. *Victorian Toronto, 1850 to 1900: Pattern and Process of Growth* 1970. 278 pp.

128. GOOD, CHARLES M. *Rural Markets and Trade in East Africa* 1970. 252 pp.

129. MEYER, DAVID R. *Spatial Variation of Black Urban Households* 1970. 127 pp.

130. GLADFELTER, BRUCE. *Meseta and Campiña Landforms in Central Spain: A Geomorphology of the Alto Henares Basin* 1971. 204 pp.

131. NEILS, ELAINE M. *Reservation to City: Indian Urbanization and Federal Relocation* 1971. 200 pp.

132. MOLINE, NORMAN T. *Mobility and the Small Town, 1900–1930* 1971. 169 pp.

133. SCHWIND, PAUL J. *Migration and Regional Development in the United States, 1950–1960* 1971. 170 pp.

134. PYLE, GERALD F. *Heart Disease, Cancer and Stroke in Chicago: A Geographical Analysis with Facilities Plans for 1980* 1971. 292 pp.

135. JOHNSON, JAMES F. *Renovated Waste Water: An Alternative Source of Municipal Water Supply in the U.S.* 1971. 155 pp.

136. BUTZER, KARL W. *Recent History of an Ethiopian Delta: The Omo River and the Level of Lake Rudolf* 1971. 184 pp.

137. HARRIS, CHAUNCY D. *Annotated World List of Selected Current Geographical Serials in English, French, and German* 3rd edition 1971. 77 pp.

138. HARRIS, CHAUNCY D., and FELLMANN, JEROME D. *International List of Geographical Serials* 2nd edition 1971. 267 pp.

139. Mc MANIS, DOUGLAS R. *European Impressions of the New England Coast, 1497–1620* 1972. 147pp.

140. COHEN, YEHOSHUA S. *Diffusion of an Innovation in an Urban System: The Spread of Planned Regional Shopping Centers in the United States, 1949-1968* 1972. 136 pp.

141. MITCHELL, NORA. *The Indian Hill-Station: Kodaikanal* 1972. 199 pp.
142. PLATT, RUTHERFORD H. *The Open Space Decision Process: Spatial Allocation of Costs and Benefits* 1972. 189 pp.
143. GOLANT, STEPHEN M. *The Residential Location and Spatial Behavior of the Elderly: A Canadian Example* 1972. 226 pp.
144. PANNELL, CLIFTON W. *T'ai-chung, T'ai-wan: Structure and Function* 1973. 200 pp.
145. LANKFORD, PHILIP M. *Regional Incomes in the United States, 1929–1967: Level, Distribution, Stability, and Growth* 1972. 137 pp.
146. FREEMAN, DONALD B. *International Trade, Migration, and Capital Flows: A Quantitative Analysis of Spatial Economic Interaction* 1973. 202 pp.
147. MYERS, SARAH K. *Language Shift Among Migrants to Lima, Peru* 1973. 204 pp.
148. JOHNSON, DOUGLAS L. *Jabal al-Akhdar, Cyrenaica: An Historical Geography of Settlement and Livelihood* 1973. 240 pp.
149. YEUNG, YUE-MAN. *National Development Policy and Urban Transformation in Singapore: A Study of Public Housing and the Marketing System* 1973. 204 pp.
150. HALL, FRED L. *Location Criteria for High Schools: Student Transportation and Racial Integration* 1973. 156 pp.
151. ROSENBERG, TERRY J. *Residence, Employment, and Mobility of Puerto Ricans in New York City* 1974. 230 pp.
152. MIKESELL, MARVIN W., editor. *Geographers Abroad: Essays on the Problems and Prospects of Research in Foreign Areas* 1973. 296 pp.
153. OSBORN, JAMES. *Area, Development Policy, and the Middle City in Malaysia* 1974. 273 pp.
154. WACHT, WALTER F. *The Domestic Air Transportation Network of the United States* 1974. 98 pp.
155. BERRY, BRIAN J. L., et al. *Land Use, Urban Form and Environmental Quality* 1974. 464 pp.
156. MITCHELL, JAMES K. *Community Response to Coastal Erosion: Individual and Collective Adjustments to Hazard on the Atlantic Shore* 1974. 209 pp.
157. COOK, GILLIAN P. *Spatial Dynamics of Business Growth in the Witwatersrand* 1975. 143 pp.
158. STARR, JOHN T., Jr. *The Evolution of Unit Train Operations in the United States: 1960–1969—A Decade of Experience* 1976. 247 pp.
159. PYLE, GERALD F. *The Spatial Dynamics of Crime* 1974. 220 pp.
160. MEYER, JUDITH W. *Diffusion of an American Montessori Education* 1975. 109 pp.
161. SCHMID, JAMES A. *Urban Vegetation: A Review and Chicago Case Study* 1975. 280 pp.
162. LAMB, RICHARD. *Metropolitan Impacts on Rural America* 1975. 210 pp.
163. FEDOR, THOMAS. *Patterns of Urban Growth in the Russian Empire during the Nineteenth Century* 1975. 275 pp.
164. HARRIS, CHAUNCY D. *Guide to Geographical Bibliographies and Reference Works in Russian or on the Soviet Union* 1975. 496 pp.
165. JONES, DONALD W. *Migration and Urban Unemployment in Dualistic Economic Development* 1975. 186 pp.
166. BEDNARZ, ROBERT S. *The Effect of Air Pollution on Property Value* 1975. 118 pp.
167. HANNEMANN, MANFRED. *The Diffusion of the Reformation in Southwestern Germany, 1518-1534* 1975. 248 pp.
168. SUBLETT, MICHAEL D. *Farmers on the Road. Interfarm Migration and the Farming of Noncontiguous Lands in Three Midwestern Townships, 1939-1969* 1975. 228 pp.
169. STETZER, DONALD FOSTER. *Special Districts in Cook County: Toward a Geography of Local Government* 1975. 189 pp.
170. EARLE, CARVILLE V. *The Evolution of a Tidewater Settlement System: All Hallow's Parish, Maryland, 1650-1783* 1975. 249 pp.
171. SPODEK, HOWARD. *Urban-Rural Integration in Regional Development: A Case Study of Saurashtra, India—1800—1960* 1976.
172. COHEN, YEHOSHUA S. and BERRY, BRIAN J. L. *Spatial Components of Manufacturing Change* 1975. 272 pp.
173. HAYES, CHARLES R. *The Dispersed City: The Case of Piedmont, North Carolina* 1976. 169 pp.
174. CARGO, DOUGLAS B. *Solid Wastes: Factors Influencing Generation Rates* 1976.
175. GILLARD, QUENTIN. *Incomes and Accessibility. Metropolitan Labor Force Participation, Commuting, and Income Differentials in the United States, 1960–1970* 1976.
176. MORGAN, DAVID J. *Patterns of Population Distribution: A Residential Preference Model and Its Dynamic* 1976.
177. STOKES, HOUSTON H.; JONES, DONALD W. and NEUBURGER, HUGH M. *Unemployment and Adjustment in the Labor Market: A Comparison between the Regional and National Responses* 1975. 135 pp.
178. PICCAGLI, GIORGIO ANTONIO. *Racial Transition in Chicago Public Schools. An Examination of the Tipping Point Hypothesis, 1963—1971* 1976.
179. HARRIS, CHAUNCY D. *Bibliography of Geography. Part I. Introduction to General Aids* 1976. 288 pp.